how to take a penalty

By the same authors and published by Robson Books:

What Is a Googly? by Rob Eastaway
Why Do Buses Come in Threes? by Rob Eastaway and Jeremy Wyndham
How Long Is a Piece of String? by Rob Eastaway and Jeremy Wyndham
Mindbenders and Brainteasers by Rob Eastaway and David Wells

how to take a penalty

The hidden mathematics of sport

ROB EASTAWAY
AND JOHN HAIGH

Illustrations by Barbara Shore

ROBSON BOOKS

First published in Great Britain in 2005 by Robson Books, The Chrysalis Building, Bramley Road, London W10 6SP

An imprint of Chrysalis Books Group plc

British Library Cataloguing in Publication Data
A catalogue record for this title is available from the British Library.

ISBN 1 86105 836 5

Illustrations by Barbara Shore

Typeset in Times by SX Composing DTP, Rayleigh, Essex
Printed by Creative Print & Design (Wales), Ebbw Vale

CONTENTS

ACKNOWLEDGMENTS

This book is the third in what could be regarded as a series. Rob wrote *Why Do Buses Come in Threes?* and *How Long Is a Piece of String?* with Jeremy Wyndham. It was a shock when Jeremy died suddenly after an operation in 2003. He always loved sport, particularly snooker and cricket, and we know he would have read this book avidly.

Jeremy would also have approved of the principle of researching material in the pub. To this end, thank you to Peter Barker and Colin Mayes for a number of ideas inspired over a glass of beer, and to Richard Harris, Chris Healey and George Westbrook, who perhaps needed something stiffer to hand when they were reading through the early drafts of the manuscript.

Throughout the book we have used and commented on findings from a host of different research papers and articles over the years. Rather than giving credit to named mathematicians and statisticians within the text, we have listed the works that we have consulted in the reference section at the end of the book. Of course all credit is due to those original researchers, and any errors are ours.

We recognise too the efforts of many individuals and organisations to make information on athletics records, football scores, and the like, easily available via the world wide web.

Many others also gave us valuable insights, including Benedict Bermange, Marc Thomas, Martin Daniels, Roger White, Tom Bramald, Tarnai Tibor and that helpful chap in the football museum whose name we forgot to note down. And we should not forget Radka Newby, who was responsible for first bringing us together in 1999 to speak at a big sixth-form event – it was backstage that we discovered our shared love of sport.

Finally, thanks to Elaine, Kay, Charlotte and the Robson team for their encouragement throughout.

INTRODUCTION

The sports world is very coy about its interest in mathematics. Commentators will play down their ability to do mental arithmetic. We even heard one *apologise* after converting the speed of a tennis serve from kilometres per hour to miles per hour! And any discussion in which the 'maths' goes deeper than that will invariably be sidelined as geeky.

And yet . . . the fact is that most participants in sport are mathematicians. They have to be, because sporting success and failure are predominantly measured using numbers, and also because many of the tactics essential to a competitor require logical, analytical thought that is, essentially, maths. It may not be maths in the form that we did it at school, but the nature of the thinking is just the same.

A cricketer standing under a steepling catch doesn't consciously solve a differential equation, but that is actually what is required in order to predict the path of the ball. And when the home team is battling to get enough points to qualify for the next round of the World Cup, the airwaves are filled with pundits and members of the public talking through the complex permutations and conditions for success.

So in this book, we will stand up and be counted. We are interested in sport, but we are also interested in mathematics, and when the two worlds meet (as they often do) the combinations can be fascinating. Sometimes mathematics can provide new insights into sporting strategies. At other times, it merely offers a deeper understanding of what we all instinctively know. And sometimes it simply throws up curiosities. A football fan once wrote to his local newspaper that 'The only point worth remembering about Port Vale's match with Hereford on Monday was the fact that the attendance figure, 2,744, was a perfect cube, 14 x 14 x 14'. No doubt that disillusioned supporter will find further nuggets in the pages that follow.

We considered carefully how much explicit 'maths' to include in the text. Although we have occasionally written down formulae, our decision was that many readers would prefer to gloss over the justifications, and read about the conclusions. But those with a taste for mathematical detail should visit the Appendix, and consult the many excellent references that we have used in compiling this book.

Our first thought was to devote each chapter to one sport. However, as we gathered more material, it became clear that certain themes cross between sports, and often the links between two sports can be stronger than those within a single sport. We didn't begin with the notion of one chapter connecting boxing to figure skating, or another linking football to golf, or snooker to rugby. But that's the way it has turned out.

This is a dip-in book, without any obvious beginning, middle or end. But if there is one theme that unifies it, it is that maths and sport are inextricably linked. For those who ask the question 'What's the relevance of maths?', we hope this book provides at least part of the answer. And we hope it is as much fun to read as it was to write.

1

BALLS

And why they aren't quite spherical

Of all the balls in all the world's sporting tournaments, there is one that stands out from the rest. Created by Adidas and named the 'Telstar', the ball was first exposed to a global audience when it was used for the Mexico World Cup of 1970. It soon became the standard design for every football, and today it is one of the most popular icons in sport.

The familiar black-and-white pattern was apparently chosen because it was more visible on a black-and-white TV than the earlier monochrome balls. However, the underlying pattern of pentagons and hexagons was not the invention of Adidas at all. A spherical object with exactly the same pattern as the Telstar football can be found on the tomb of Sir Anthony Ashley in the parish church of Wimborne St Giles in Dorset, dating to the mid 1600s. Some have suggested it might be England's oldest football. Its resemblance to a football is enhanced by the fact that the ball is to be found close to Sir Anthony's feet, though historians believe it is more likely to be some sort of heraldic symbol marking his work as a navigator. Sketches of the shape go back even further, to

Leonardo da Vinci, and the principle was certainly understood in the time of Archimedes.

The Telstar ball is made up of 32 panels. Twelve of them are identical regular pentagons, coloured black, and the other 20 are white hexagons. (Cartoonists who don't work from a real model often get this wrong, drawing a pattern of black hexagons instead of pentagons.)

Why did Adidas go for this combination of geometrical shapes? The answer is that apart from having a striking appearance, this is a very effective way of creating a nearly spherical object from flat panels. Try putting together a 'ball' with any other combination of regular shapes and the result will have unsatisfactory bulges and points that will wreck the aerodynamics of the ball, even though they are smoothed out to some extent once the ball has been inflated.

This classic football shape has a formal name. It is a *truncated icosahedron*, but it is also known informally as a buckyball, named after the architect Buckminster Fuller, who invented a strong and highly efficient form of structure known as a geodesic dome. You will find such domes all over the world, for example at the Epcot Centre in Disney World and the Eden Project in Cornwall.

Platonic balls

Life would be much easier for the manufacturer of footballs if all the flat panels were the same, regular shape. In fact, as the ancient Greeks knew, only *five* such designs are possible. These are the so-called 'Platonic solids', made up of triangles, squares or pentagons, and if their faces were made of a flexible enough material, the flat-faced solids could be inflated into a spherical shape:

- The tetrahedron – four faces ('hedron' means face), each an equilateral triangle

The tetra-ball

- The cube – the familiar die, with six squares as its faces

The cubo-ball

- The octahedron, or diamond – eight faces, each an equilateral triangle, rather like two square pyramids whose bases are glued together

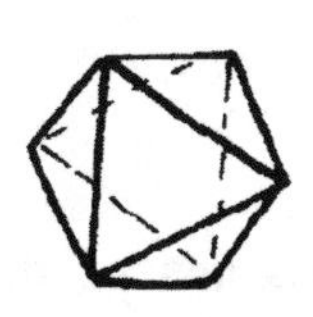

The octa-ball

- The dodecahedron – twelve faces, each a regular pentagon

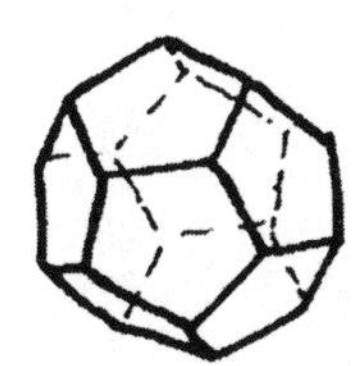

The dodeca-ball

- The icosahedron – twenty faces, each an equilateral triangle.

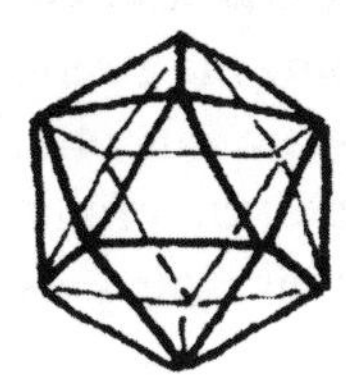
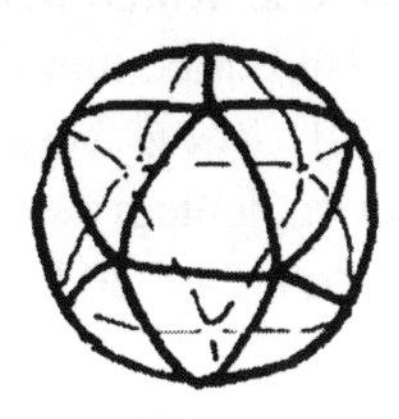

The icosa-ball

And that is it: there are no more solids that are made up entirely of identical, equal-sided shapes.

The first three 'balls' are too far from a sphere to be practical. Even when inflated to make the sides curved, they would still have points and ridges that would give the balls unpredictable bounce and flight (something that goalkeepers complain about even with near-spherical balls).

The dodecahedron and the icosahedron do both make fairly reasonable spheres, but still neither is close enough for practical use. The icosahedron has the added disadvantage that five panels would need to be stitched together at each vertex, a fiddly task that wouldn't be appreciated on the factory floor.

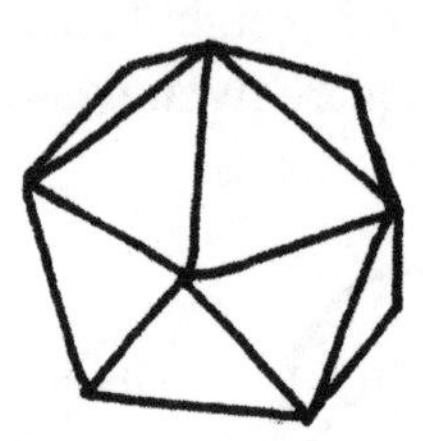

However, there is a simple way of solving the problems of the icosahedron. Just snip off the vertices, like this:

Only three panels now meet at each vertex, which makes it much easier to stitch the panels together. You will also notice that lopping off a vertex creates a face that is a pentagon. If all twelve vertices are cut off in the same way, you end up with twelve pentagons, while the original triangular faces have all been cut down to create 20 hexagons, with every edge the same length.

This is the truncated icosahedron that dominates world football. And it comes about as close to a sphere as you can get for a solid made up of regularly shaped panels.

As it happens, you can make the truncated icosahedron even more spherical if you make the pentagons slightly larger than in the perfectly regular version, and this is the method used in the manufacture of some of the better quality football brands. Inspect a modern ball and you may find that the hexagons are not quite regular: the sides that are shared with pentagons are marginally longer than the others. (Of course the hexagons may also be irregular because the ball is low quality – especially if it is a cheap plastic ball where the shapes have been painted on.)

If FIFA wanted to come up with a pretty new design for a football, there is another shape they could consider. It has 62 faces, of which 20 are triangles, 30 are squares and 12 are pentagons. It is a little more spherical than a truncated icosahedron, and each of the different shapes could have a different colour.

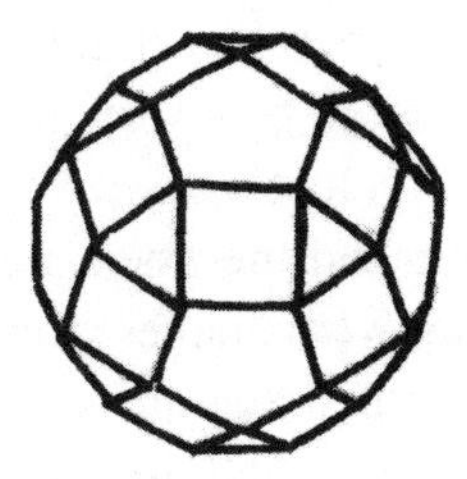

Its mathematical name is a *rhombicosidodecahedron*. One disadvantage is that it would take up most of the half-time analysis just to say it.

Golf ball dimples

Footballs aren't the only balls on whose outer surface hexagons and pentagons are used. If you examine a new golf ball closely, you will see that the small dimples all have these shapes. Once again, these shapes are chosen because hexagons and pentagons are a good way of covering a sphere, though in the case of a golf ball they also serve another purpose. The secret of a good golf ball is in the aerodynamics. If a golf ball were perfectly smooth then even Tiger Woods or John Daly would only be able to hit it about half the distance they currently achieve.

The idea of a dimpled golf ball was discovered by accident. In the mid 1800s, golf balls were made from the hardened rubber-like sap of the gutta tree, and were sold perfectly smooth. After a while, golfers noticed that old balls travelled further than new ones, the only difference between the two being that the old balls had acquired pits and scratches after so much use. Ball-makers therefore began to insert indented patterns into the balls to replicate the effect of wear, and after much trial and error, the modern style of golf ball with its regular dimples evolved.

The effect of the dimples is to reduce the amount of drag on the ball. They do this by creating a turbulent band of air around the surface of the ball. The uncomfortable experience of turbulence when sat in an aircraft might suggest that turbulence is a bad thing, but when it comes to flow over a moving ball, it turns out to be a very good thing. It forces the air to hug the surface of the ball closely, and (engineers have discovered) reduces the size of the wake and the associated drag force. By the way, this same principle explains what makes a cricket ball swing. Bowlers polish only one side of the ball so that one side is smooth, while the other is left rough, and this creates turbulent drag that causes the ball to swerve.

There are ways of demonstrating all this mathematically, but not in a single paragraph, so we'll simply pass on the observation that a cricket ball can swerve alarmingly, and a cunningly dimpled golf ball travels twice as far as a smooth one.

Until recent years, the dimples on golf balls were usually circular. The disadvantage with this was that between the circles there were sizable flat areas that didn't create much air turbulence. The introduction of hexagon–pentagon patterns has meant that the flat regions between the dimples are now much smaller, and this has added yards to the distance golfers can hit the ball.

The technology of dimple design has become extremely sophisticated, with different manufacturers adopting different strategies. On some balls, the hexagons vary in size – the large ones are said to reduce drag, while the small ones stabilise the ball in flight.

Other ball patterns

Not all sports balls use the hexagon–pentagon principle. The most primitive design of ball has the appearance of a peeled orange, with the seams running from the top of the ball to the bottom, all meeting at what could be called the north and south poles.

Early footballs had this segment design, and simple juggling balls still do. It was probably the inspiration for the eight-panel basketball patented in 1929 by G L Pierce, and which is used to this day. The disadvantage of the orange-segment design is that all the panels meet at a single point, which creates an unsatisfactory lump. This is probably why Pierce's basketball pattern added curves to the panels so that they join like this:

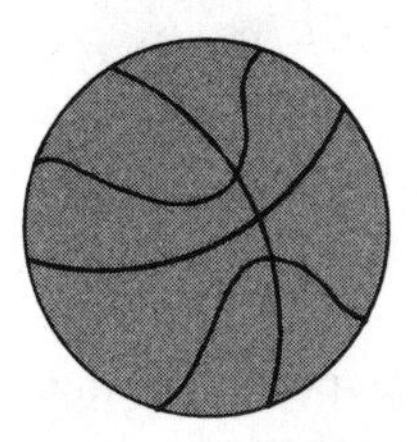

Basketball quiz

One of the diagrams below represents the silhouette of a basketball as it is passing through a regular basketball hoop. Which is the correct diagram?

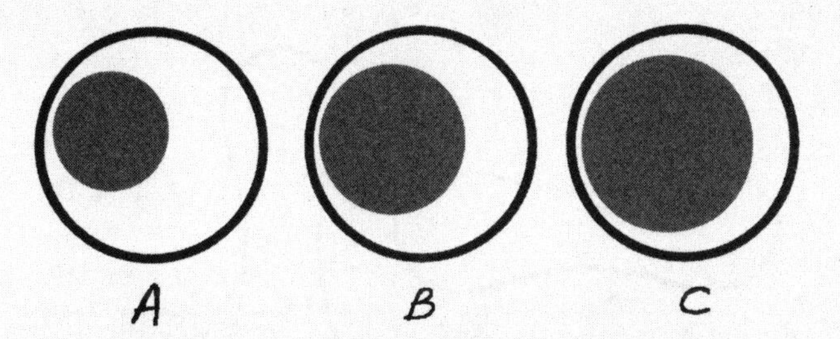

(See the end of the chapter for the answer.)

One of the most aesthetically pleasing balls to look at is the tennis ball. A pleasant curve wends its way around the circumference, dividing the ball's surface into two identical regions.

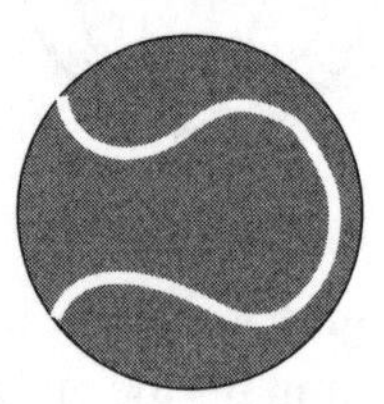

Viewed from one direction, the tennis ball in fact resembles a yin and yang symbol:

This design of tennis ball replaced the orange-segment style in the 1870s, but tennis almost certainly pinched the idea from baseball, which used this style of ball from the 1840s onwards.

The inspiration for the design was just the same as for the modern football, that is, to make a ball that was as spherical as possible, while at the same time keeping the materials simple and the amount of stitching to a minimum. The inventors of the baseball discovered that something close to a sphere could be created using two identical dog-bone shaped pieces of leather:

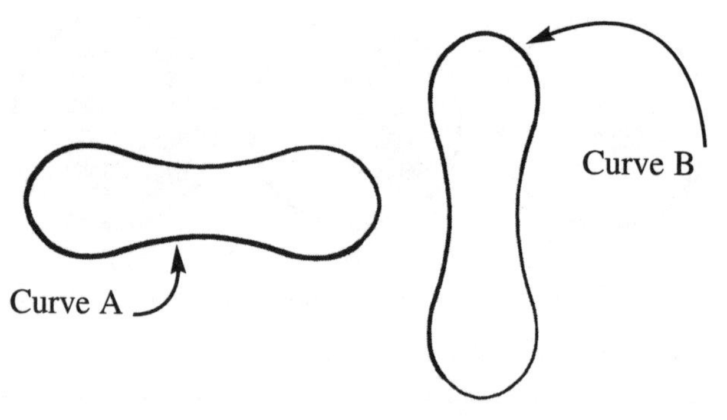

These two pieces can be stitched together, curve A fitting snugly with curve B. (Cut apart a baseball or a tennis ball along the seam and you will discover two pieces shaped like this.)

What wasn't clear to the inventors was which shape of dog bone to choose. You can probably imagine that the two pieces would fit together if they had a dumbbell appearance with very narrow necks and bulbous ends, like this:

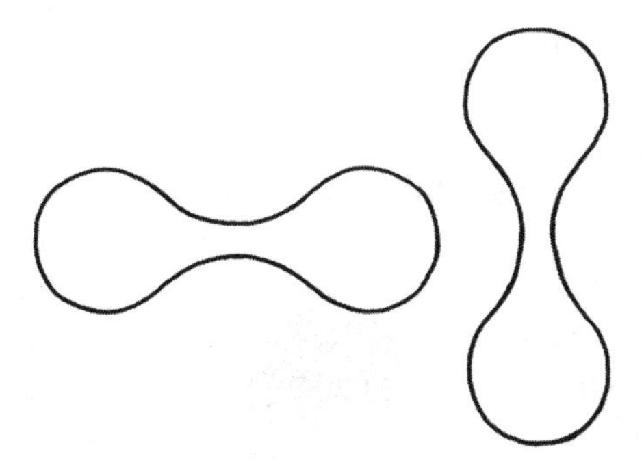

The thin necks will be nicely curved, but the round ends will be rather flat.

At the other extreme, you could have two rectangles, like this:

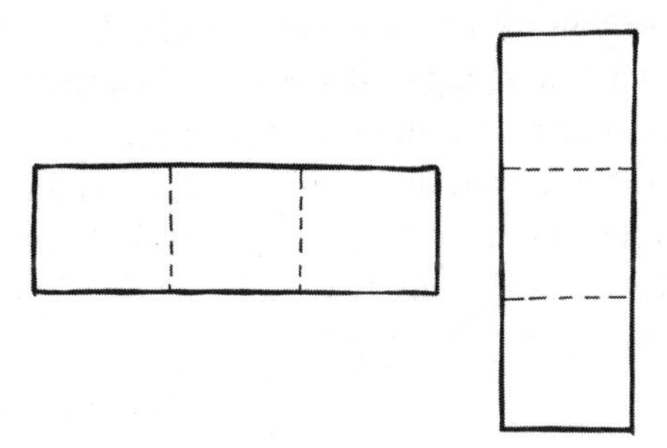

In this case, the two pieces stitch together to make a cube, which, as we saw earlier, is not exactly ideal as a baseball!

The best ball must lie somewhere between the two, but complex geometry probably wasn't the strong suit of these early designers so they used the next best thing – trial and error. In fact the design for the pieces that go together to make a baseball was influenced by how easy it was to grip the stitched ball. Whatever the reasoning, by luck or by design the shape of the leather panels used to make a baseball turned out to be almost the perfect choice for creating a spherical ball.

A tennis ball may look as if it is made the same way as a baseball, but in fact the inside is a rubber ball made of two hemispheres. The dog-bone shaped yellow parts on the outside are made of flexible material, and simply glued on. For this reason, the curve of the shapes is less critical, and in fact the neck of the bone shape on the modern tennis ball is wider than that on the baseball.

From tennis to Vennis

Tennis balls have inspired at least two quirky mathematical ideas.

The first is that if you have a ball with hairs that stick out of it – a tennis ball, for example – and try to comb all of the hairs flat, mathematicians have shown it just can't be done. There will always be ridges where the hairs ride up against each other, no matter how you try. You will find references to the Hairy Ball Theorem in books about topology. (One consequence of this theorem is that, at any given instant, there must be at least one spot on the earth's surface where the air is completely still. If you picture the earth as the ball and the local wind direction as the 'hairs' you might see the equivalence.)

The second tennis ball idea relates to Venn diagrams, which you might recall from schooldays as a way of illustrating overlapping sets. It is easy to draw three overlapping sets on a piece of paper:

Venn diagram

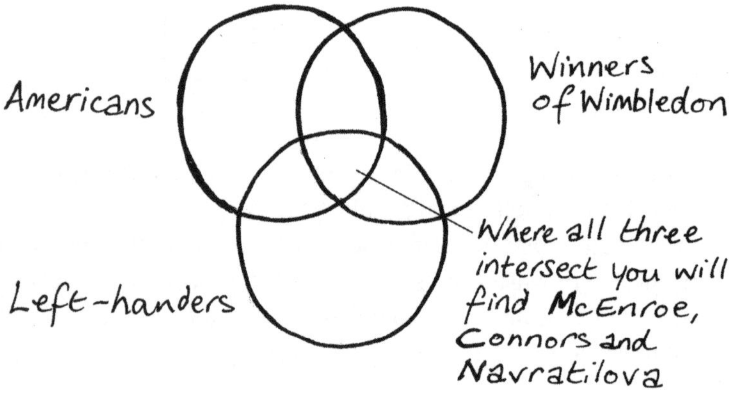

However, suppose you want to add a fourth category to your Venn diagram. The picture above becomes much messier if you attempt to add another category that intersects with every combination of the other three. Let's suppose we want to add the set of 'Tennis players who have presented TV quiz programmes' to the diagram.

A tennis ball can help. The curved line divides the ball into two regions, which you can label Americans and non-Americans.

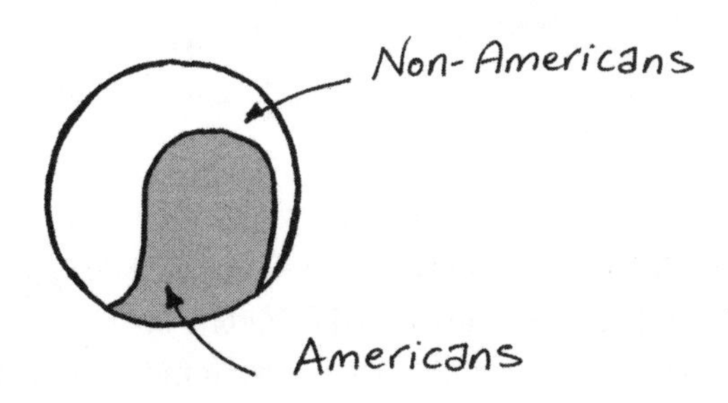

Now mark an equator that cuts the seam at four equally spaced places. (There is only one place you can do this.) This divides Wimbledon winners from non-winners.

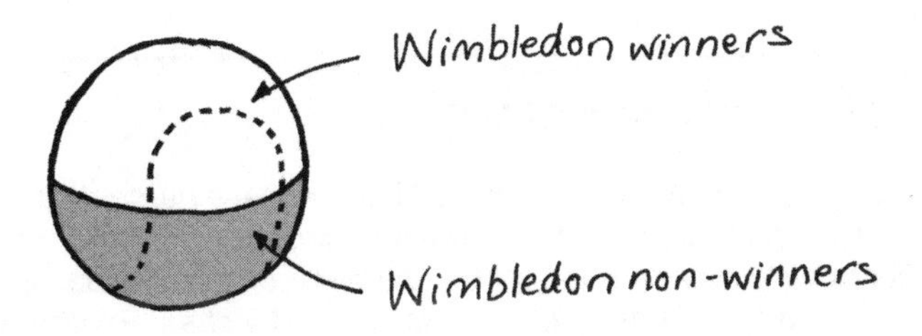

Now locate the north and south poles, and join them with a complete circle down the middle of one of the curved strips – call this the Greenwich Meridian. This divides left- and right-handers.

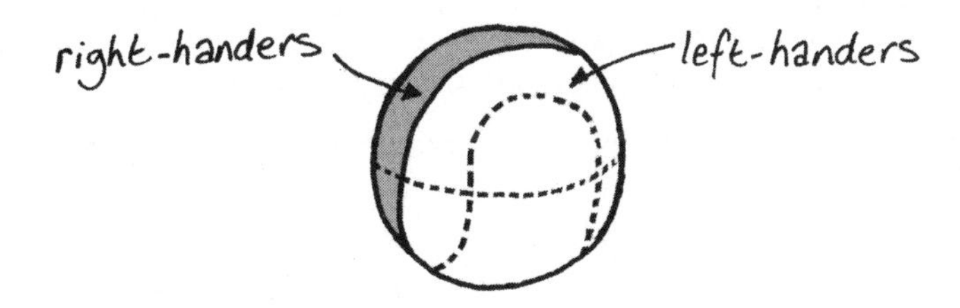

Finally, draw another complete circle through the poles, 90 degrees away. Use this to divide TV quiz presenters from the rest.

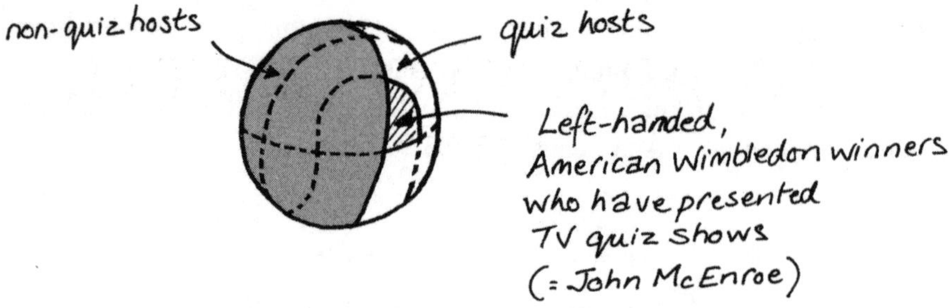

non-quiz hosts

quiz hosts

Left-handed,
American Wimbledon winners
who have presented
TV quiz shows
(= John McEnroe)

You now have a Venn diagram that represents all the regions for four sets. For this reason, it has been christened a *Vennis ball.*

The region on the Vennis ball that represents American left-handers who have won Wimbledon and presented TV quizzes is indicated on the diagram. We think it has only one member: John McEnroe.

Might you ever have practical need to use the 'Vennis ball'? You could always demonstrate the principle to your opponent during the change of ends. The mental distraction might just put them off their serve.

Answer to the basketball hoop quiz

The answer may surprise you – it is A. Most people's perception is that a basketball fits quite tightly into the basket, and they therefore choose answer B or C. But the truth is that the diameter of a basketball is about 9.5 inches, while the diameter of the basket is 18 inches. In other words, you could almost squeeze two basketballs through at the same time.

The reason for the so-called 'basketball hoop illusion' is that people have a tendency to underestimate the dimensions of elevated objects that they never get to see up close. The same principle applies to road signs and chimneypots. Perhaps the basketball illusion is reinforced by the fact that it also seems so difficult to get the ball through the hoop; we assume this is due to the small size of the target.

2

WHY DIDN'T YOU BELT IT, SON?

When sport meets game theory

When the Euro 1996 semi-final between England and Germany ended in a draw, the match had to be resolved by penalties. With the scores balanced at a nail-biting 5–5, Gareth Southgate stepped up to take the next kick. He tried to place it carefully, the keeper dived the right way and saved easily. The next penalty from Germany was slotted home, and England were out.

'Why didn't you just belt it, son?' asked Southgate's mum after the game. Mrs Southgate had obviously forgotten that, six years earlier, Chris Waddle of England had 'belted' the ball yards over the crossbar in a similar shoot-out with West Germany during the 1990 World Cup. Hindsight is a wonderful gift.

But what *should* a penalty taker do? Should he belt it or place it? Should he aim left or right? It all looks so simple when you watch it on TV, but anyone who has had to suffer the mental pressure of taking a penalty will realise that penalty taking is far from a trivial exercise. In fact, it is a very real application of the deceptively complex world of *game theory*.

Game theory is all about making decisions. More precisely, it is how to maximise your chance of winning when you have to take into account what your competitors are thinking. It is the world of bluff and double-bluff. It applies in business, where two companies try to anticipate each other's advertising campaign, and in economics where it has been the topic of Nobel prize-winners. It applies in war, when a general tries to bluff the enemy into sending their troops the wrong way. It applies in the natural world, when a gazelle tries to sell a dummy to an attacking lion.

It applies in the popular children's game of paper-scissors-stone. And it applies when taking penalties.

The choice of tactics

The 'game' of penalty taking is a simple one. The kicker is trying to score, while the goalkeeper, of course, aims to keep the ball out. Only one of the two can be successful, so this is a win–lose game, or, if you want to impress friends with a technical definition, it is a *zero-sum two-person game*.

Both of the players in this game have different strategies they can use. The kicker can aim left, right or centre, high or low, and he can blast it or place it, or gently chip it. He might even follow the audacious lead of Antonin Panenka, whose gentle dink into the empty centre of the goal won the penalty shoot-out for Czechoslovakia in Euro 1976. When it works, it smacks of gutsy impudence or even arrogance. But a soft kick that doesn't work results in humiliation, as Gareth Southgate discovered to his cost.

The goalie, meanwhile, can make a premeditated decision to jump left or right, he can attempt to read the kicker's body language, or he can try to follow the ball after it has left the foot, in the hope that he can deflect it before it crosses the goal line. There are, in short, plenty of choices. The question is, which is best for the kicker, and which is best for the goalkeeper?

It is easier to get an idea of how penalty strategy works by simplifying the picture. Let's assume that a penalty taker (call him Beckham) has only two choices:

- He attempts to place the ball in the corner of the net
- Or . . . he belts it somewhere near the centre of the goal.

Meanwhile, we'll assume the keeper has a choice of just two strategies:

- He makes a premeditated dive left or right
- Or . . . he stands still.

In this simplified situation, there are four possible scenarios, and each of them is likely to have a different probability of ending with a goal. For example, suppose Beckham chooses to aim straight. If the goalie happens to have chosen to stand still, then the chances are high that he'll save it. A sensible estimate for the chance of a goal in this situation would be 30 per cent:

Beckham's choice	What the goalkeeper does	
	Stand still	Dive to one corner
Aim straight	30%	
Aim at a corner		

The rest of the grid can be filled in the same way. If Beckham aims straight and the keeper dives, there's a high chance, call it 90 per cent, of a goal. If Beckham aims for the corner and the keeper stands still, the chance of a goal is still high, though maybe with a slight risk of shooting wide, so we'll call it 80 per cent. And if Beckham aims for a corner and the goalie dives, the chance of a goal might be only 50 per cent. The completed grid looks like this:

Beckham's choice	What the goalkeeper does	
	Stand still	Dive to one corner
Aim straight	30%	90%
Aim at a corner	80%	50%

Remember this is only a simplified model of what really goes on, in order to illustrate how game theory works in this case.

Given these probabilities, our theoretical Beckham is faced with a dilemma. What would you do if you were him?

If the keeper stands still, Beckham's best chance of scoring is to aim for the corner, but if the keeper dives, Beckham does better if he simply blasts it straight. So Beckham would like to know what the keeper is going to do.

Meanwhile, the keeper knows that his best chance is if Beckham kicks it straight and he stands still, since then a goal will be scored only 30 per cent of the time. However, if Beckham realises the keeper is going to stand still, then he will aim at the corner, with an 80 per cent chance of a goal.

This is the heart of the penalty taker's dilemma. No single strategy will deliver the best chance of a goal: if Beckham always aims straight, his chance of scoring will only be 30 per cent because the keeper will decide always to stand still; and even if Beckham always aims at the corner, his chance of a goal will increase only to 50 per cent because the keeper will always dive.

With the above probabilities, is there a strategy for a striker to use that guarantees him a better chance than 50 per cent of scoring? As it happens, there is.

Being unpredictable

In the game of penalty taking, as in many other games, the key is in not being predictable. Beckham needs to mix his strategies between aiming straight and placing the ball in the corner, but he needs to do so *randomly.*

Randomness has a particular mathematical meaning. Something is random if what has happened before has no bearing on what will happen next. The toss of a coin (as we will discuss in Chapter 6) is random because a sequence of ten heads doesn't alter the fact that the chance of the next toss ending up as a head is still 50–50. The roll of a dice is also random, with the chance of rolling a six always being 1 in 6, whatever happened last time. On the other hand, kicking the ball left because you kicked it right last time is definitely *not* random.

It isn't obvious how Beckham should divide his kicks between placing in the corner and aiming straight. Should he belt it straight 50 per cent of the time and place it the other 50 per cent? In the particular example above, if he were to divide his kicks 50–50, then it turns out that (against a knowledgeable keeper) he can expect to score about 55 per cent of the time. However, he can improve his chances further – to a guaranteed 63.3 per cent, in fact – if he chooses to blast the ball *exactly one-third* of the time and to place it on all other occasions. If you want to know where these figures come from, look in the Appendix.

How, in real life, can you make such a random choice? We will see in Chapter 6 that people are notoriously bad at 'making up' something random, so a physical aid is needed. If Beckham had to make a 50–50 choice, he could always secretly toss a coin before he took the penalty. ('Heads I place, tails I hit it straight.') However, this doesn't work if the split has to be in the ratio 2 to 1, or any other unequal division. Indeed, most of the time when such a choice has to be made, there will need to be a different chance of the two alternatives occurring. But there are plenty of ways to achieve this, if you use a bit of imagination.

We have said that in our example, Beckham needs to blast the ball one-third of the time, on average. One method he could use would be to glance at the stadium clock at the instant the ball is placed on the spot. The position of the second hand serves as a very effective random device – at any instant in the game, it will point in a completely random direction.

One-third of the time it will be between the 12.00 (midday) position and 20 minutes past the hour; the rest of the time it will be after the 20-minute spot. So if the second hand is in that first segment, he should blast the ball. And if it is after the 20-minute position, but not back to 12.00, he should place it. That gives Beckham a random choice, with the correct one-third/two-thirds frequencies. On average, if he always uses this tactic, he will blast the ball one-third of the time and place it the rest of the time, though each individual kick will be random just as the likelihood of a head is when tossing a coin.

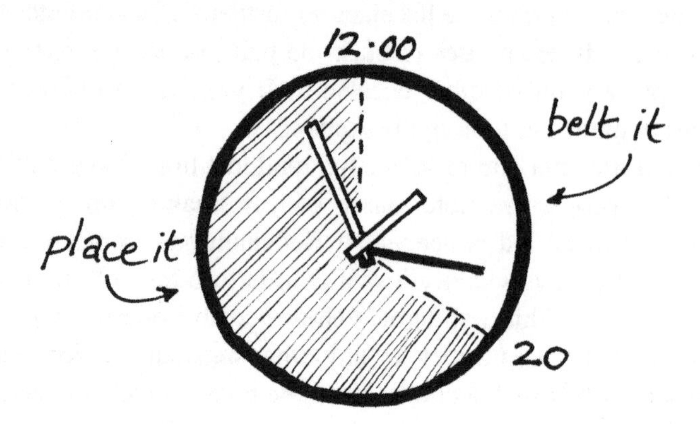

There are other ways of randomly coming up with a 1 to 2 split on the football field. For example, he could look to see which of his nine outfield team-mates is closest to the referee as he walks back: if it is one of the three defenders, he should blast the ball (one-third chance); if it is one of the six other players he should place it (two-thirds chance).

Meanwhile, the goalkeeper too should be making a random decision. Using the probabilities we have estimated, it turns out that the goalkeeper needs to stand still four-ninths of the time and dive five-ninths of the time if he is to maximise his chance of preventing a goal. He too has a variety of ways of making his choice; all that matters is that neither player should give any clue to the other as to how their choice is made – if one player leaks any information that gives an idea as to what he will do, then the odds will shift in the opponent's favour.

Back to reality

The biggest simplification in this example has been allowing each player a choice between two tactics only. In reality, the penalty taker could go for any combination of direction, height, power, etc., having a dozen or more choices. Similarly, the keeper has a long list of options. However, the principle remains unchanged: there will be a *payoff matrix*, showing the chances of a goal for each pair of choices made by the two players; there *is* a systematic approach to working out the optimum tactics for each player; and both need some practical way – the *randomising device* – to accomplish their optimal choice of action.

But is there any practical relevance for all this in the real world of penalty taking? Arguably there is, even though the probabilities in real life will never be as definite as the ones we have used in our example.

In Euro 2004, England were leading France by a goal to nil when Wayne Rooney was brought down in the box. Beckham took the penalty and Barthez saved it. These things happen, but the result was that the press began to theorise about whether Beckham had lost the knack. 'He always aims left, he should vary his direction,' was one theory. The trouble was, if Beckham listened to this advice, then what should he do next time he had to take a penalty? If he aimed left and missed, he'd be pilloried for not 'learning from his mistakes'. If he aimed right, everyone would say he wasn't his own man, he'd just succumbed to pressure and followed the advice of the media. If we take our simple model as good enough, he could simply announce that, each time he took a penalty, he would ignore the past, and give himself a one in three chance of blasting the ball straight and a two-thirds chance of aiming for one corner or the other. But HOW he made his choice at any time must remain a complete secret.

Meanwhile, what actually happened after the Barthez save? In the quarter-final against Portugal, Beckham lined up to take another penalty; he tried to belt it and famously blasted the ball several feet over the bar. If only he'd looked at the clock . . .

Factors that can change the goalkeeper's tactics

Penalty takers have been known to practise one particular kick in preparation for a game (for example, aiming for the top left corner). So long as the kick being practised remains secret, this won't affect our game theory analysis. But what if the kicker has to take a second penalty during the match? If he has concentrated on one kick, he now has the choice of repeating that kick or attempting a different, less practised and therefore less reliable kick. When deciding what to do, the goalkeeper should bear this in mind. The goalkeeper should behave randomly, but with a slightly increased chance of diving in the direction that the first penalty was kicked.

Another tip for goalkeepers is to watch the feet of the kicker just before the ball is struck. One study discovered that 85 per cent of the time the penalty taker's non-kicking foot points in the direction he is planning to kick the ball. If the goalkeeper can react quickly enough to this clue, his odds will improve.

Team selection and the Ryder Cup

Penalty taking is probably the most common and explicit situation in sport where bluff and counter-bluff game theory comes into play, but it's not the only example. Every time a player sells a dummy, or disguises a drop shot, he's using a bit of ad hoc game theory. An in-depth analysis of fluid situations like these would be pushing credibility, since every situation is different, but there are a few 'set pieces' where game theory has interesting things to say.

One example is in team selection. In football, as in many sports, teams are not usually announced until just before the start of play. Neither manager knows for sure which team the opponents will put out, and since the tactics of the game are partly dependent on knowing exactly who the opposition will be, this presents something of a dilemma.

One of the neatest examples comes not in football but in golf's Ryder Cup. On the final, critical day of this tournament, the American and European captains have to announce the order in which their twelve players will be going out onto the course. In each of the twelve matches, the winner gains a point for his team.

A captain might choose to send out his strong players first (to boost morale when the weaker players follow) or he might do the complete opposite (to make sure he has strong players at the finish when the pressure is greatest). On top of this, however, the captain knows that putting his very strongest player against the very weakest of the opposition might be a waste. A point is a point whether you win by ten holes or by a single putt. Why put Tiger Woods in his pomp up against one of Europe's lesser players when he might be more useful to his team if he took on one of Europe's strongest players, leaving a weaker US player to take the point from the lowest-ranked European?

But the tactics of who to play against whom might also be influenced by the state of the match. To keep the numbers manageable, let's imagine two teams with five players each. Also imagine that, overall, the Americans have a marginally stronger team: among the ten players, they are ranked 1, 3, 5, 7 and 9, while the Europeans are ranked 2, 4, 6, 8 and 10. We offer three possible sets of pairings.

In Draw A, the USA has the stronger player in each match, but only just.

DRAW A

USA	Europe
(rank of player)	
1	2
3	4
5	6
7	8
9	10

In Draw B, the USA has the stronger player in the first three matches, while Europe are favourites in the other two.

DRAW B

USA	Europe
1	10
3	8
5	6
7	4
9	2

Finally, in Draw C, Europe are ranked higher in four of the five matches (this is the best situation they could hope for).

DRAW C

USA	Europe
1	10
3	2
5	4
7	6
9	8

Three draws have produced three very different looking scenarios. Will these influence the result? On average, not by much. In each of the draws above the USA might expect to win by roughly 3 matches to 2 on average. However, the range of likely outcomes is wider with Draw A than the other two. We can see why just by looking at the first match in each draw. In Draws B and C, the USA is almost certain to win the first match, because the best player is paired against Europe's worst. So if the USA needed only one point to win the Ryder Cup, Draw B or C would suit them very nicely, whereas in Draw A there would be a greater risk that they might lose all five matches. In Draw C, on the other hand, Europe has a better than even chance in four matches, so if Europe wanted 4 points from these matches to win the Cup, this draw would suit them better than the other two.

Given this knowledge, is there anything that either captain can do to bend the pairings in their favour?

As is always the case in game theory, knowing the tactics of the opposition would be of enormous help. If the USA had a stated policy of putting their best players out first, then Europe would be able to dictate the pairings in every single match, aiming to give themselves the best chance of reaching whatever target score was needed to win the Ryder Cup. For instance, if only one point is required, the best tactics might be to reverse the order and put the weakest out first, expecting guaranteed points later on.

But there is a simple way to scupper wily opponents, provided your locker room is spy-proof. The strategy to use is to draw the names out of a hat, in completely random order. By doing this you prevent your opponents from being able to influence the pairings in any way at all. It doesn't even matter if they know that this is your strategy, as long as they don't get to see the actual order of the players.

Once again, randomness has come to the rescue. And in this case, no surreptitious glance at the clock is required.

3

A MATTER OF PERSPECTIVE

Spectators and referees see things differently

In 1973, the Barbarians took on the mighty New Zealand All Blacks in one of the most pulsating and physical rugby matches of all time.

There is one moment from that match that has lived on in rugby folklore, a try that swept the full length of the pitch and involved seven players. As the full-back Phil Bennett desperately ran back towards his own line to retrieve the bobbling ball, Cliff Morgan began his memorable burst of television commentary:

> *This is great stuff . . . Phil Bennett covering, chased by Alistair Scown . . .* [Bennett sidesteps his way out of trouble] . . . *Brilliant, oh that's brilliant . . . John Williams . . . Pullin . . . John Dawes . . . great dummy! To David – Tom David. The halfway line. Brilliant by Quinnell. This is Gareth Edwards . . . a dramatic start . . . what a score!!!!!*

For those of us who have seen the try, those words still bring a tingle to the spine.

There is, however, one small but crucial point that lovers of this try conveniently gloss over. If the referee had strictly applied rugby's Law 12 (and had had the benefit of an aerial view) then the try would have been disallowed, because it involved at least one 'forward pass'. And this is not the only famous try that has broken the rules. Perhaps as many as half of the great running tries should, strictly speaking, have been disallowed for the same reason. Here's why.

Forward pass confusion

One of the important rules in rugby concerns how the ball can be passed. A forward pass (officially entitled a 'throw-forward') is not allowed, which is why in attacking moves the player receiving the ball is always level with or slightly behind the one who throws it. This simple law ensures that rugby players remain in orderly formations, instead of being scattered across the field as they are in football.

Law 12 tries to make clear what constitutes a throw-forward. It says:

DEFINITION – THROW-FORWARD
A throw-forward occurs when a player throws or passes the ball forward. 'Forward' means towards the opposing team's dead ball line.

Now consider what happens in a typical attacking move.

Smith is running with the ball, and wants to pass it to Jones. Both players are racing towards the opposition try-line. Viewed from the blimp hovering above the ground, it is clear that Jones is further back than Smith, so the direction in which Smith points the ball is indeed away from the opposing team's dead ball line, as the rugby law requires.

Moments later Jones has caught the ball, and he races to the line to score a thrilling try. All seems to be in order, until you consider what must have happened to the ball while it was in flight. Jones was running full pelt, so while the pass was floating in the air towards him, he ran forward, perhaps by as much as a couple of metres. For the pass to be successful, it needed to land in his hands, not two metres behind him. This means that the actual path of the ball must have looked like this:

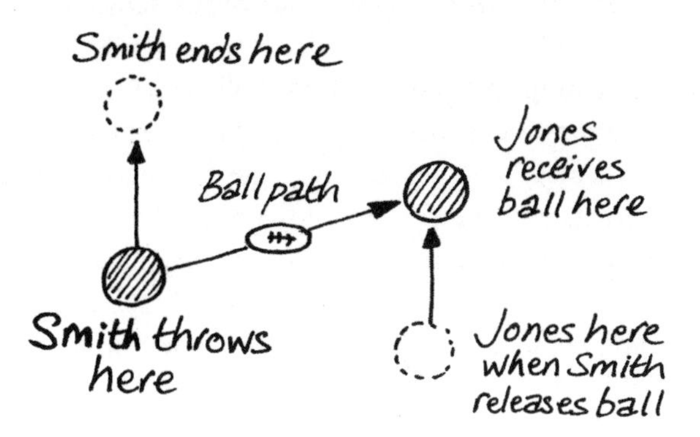

From our blimp we can now clearly see that the ball has gone forward, even though Jones was behind Smith at all times!

So how, if Smith threw the ball backwards, can it have gone forwards? The explanation is that the ball DID go backwards *relative to Smith.* One way to demonstrate what is going on is to represent the different motions as arrows (or more formally, vectors). The forward motion of Smith is represented by the arrow on the left in the diagram below. The 'backward' throw by Smith is the arrow pointing down to the right. When these two vectors are combined, they give the forward pass we saw above.

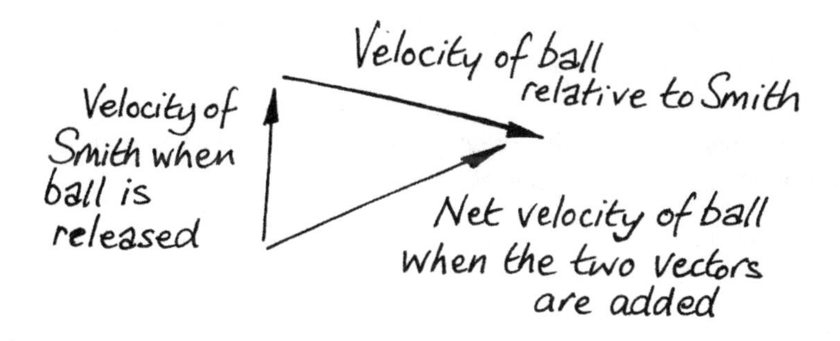

To a television viewer, seeing the game from a camera that is level with the action, this forward motion of the ball is often quite obvious (especially if the pass happens close to a line on the pitch so that its motion relative to the ground is clear). But as long as the passer keeps running forward, the touch judge, who is typically running alongside the players, is unlikely to raise his flag. As far as the sprinting touch judge is concerned, the ball has gone backwards relative to him, so it appears legal.

There is an interesting exception. Often a player will pass the ball just before he is the victim of a crunching tackle, which stops him stone dead. By the time the receiver catches the ball, he is clearly forward of the player who passed it, and the referee blows the whistle. This is clearly unsatisfactory, since the same pass would typically have been permitted if only the passer had been able to keep running. Maybe there's a lesson in this. If you are about to be tackled, don't attempt to pass the ball sideways.

Changing the law?

Some people feel that rugby's Law 12 needs to be changed to remove these obvious flaws. One suggestion for the rewording goes as follows:

> *At the moment the ball is released, the player who receives the ball should be behind the player who passes.*

After all, this is the rule for kicking in rugby, so why not for passing, too? (Recall too the offside law in soccer: a player cannot be offside if he is behind the ball when it is played.) It would, however, allow the ball to be deliberately thrown forward. Suppose the passer is standing still. Under this new wording, he could now lob the ball forwards over the defenders, allowing a winger who is behind him to hurtle at full speed and catch it well in front. This would change the whole dynamics of rugby, and would surely not be acceptable.

The way that most people *interpret* Law 12 suggests that it should actually be reworded something like this:

> *At the moment the ball is released, the passer should project the ball backwards relative to his own motion at the time.*

This would allow the ball to travel forwards relative to the passer, if he gets tackled just after release, which is theoretically fine, but would look wrong to referees and spectators.

In order to maintain the status quo, we need a law that mimics what actually goes on in real games. It appears that the rule as referees apply it is that the ball should not BLATANTLY be passed forward. (Anything for an easy life and not too many stoppages.) So in this spirit, we suggest the following rewording of Law 12:

At the moment when he catches the ball, the receiver should be level with or behind the player who released the ball.

We concede that this still creates problems, because it doesn't cover the situation where the passed ball bounces and bobbles forwards. But at least if this version of the Law were applied retrospectively, it would confirm the legality of the great Barbarians try, with the full support of millions of rugby fans.

Challenging an LBW myth

Forward passes aren't the only example of adjudicators fudging the geometrical truth for an easy life. A similar blind eye is turned when applying the LBW law in cricket.

LBW, which stands for leg-before-wicket, is one of the most contentious regulations in any sport, partly because it relies so heavily on the opinion of the umpire. The aim of this law is straightforward enough. Cricket should be about batsmen protecting their wicket with a bat. If they just padded themselves up like Michelin men and stood in front of the stumps, it would become an exceptionally tedious game, even for those of us who currently find it fascinating. This is why a law was introduced to penalise a batsman if he prevented the ball from hitting the stumps with anything other than his bat. (While this law is known as leg-before-wicket, it equally applies to arm-before-wicket, bum-before-wicket, and even head-before-wicket if the batsman ducks too low.)

The devil, however, is in the detail. For all sorts of reasons, LBW has developed into a regulation filled with conditional ifs and buts. You might wish to take a deep breath before embarking on the following description of the rule.

The convoluted LBW law

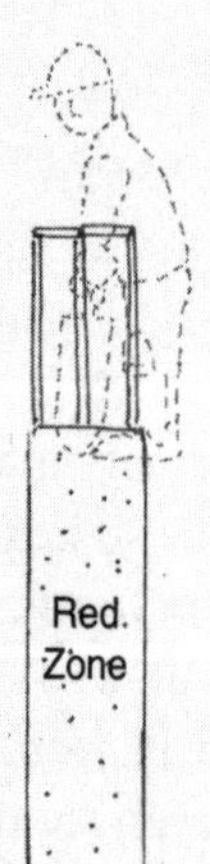

We will adopt a TV convention, and describe the narrow strip of ground between the two sets of stumps as the 'red zone'. The description we give is from the perspective of the umpire, and applies to right-handed batsmen. Make your own changes for left-handers.

If the ball hits the batsman's leg (or any other body part except the glove) and would otherwise have hit the stumps, then it is LBW if . . .

1. *The ball struck the batsman within the red zone. Or . . .*
2. *The ball struck the batsman to the left of the red zone (the off side) and the batsman did not attempt to hit it.*

However, if the ball pitched to the right of the red zone (the on side), the batsman cannot be out LBW.

. . . OK, it's over, you can relax.

What it means is that when the ball strikes the batsman's pads, the umpire has to make a series of judgments about the trajectory of the ball, where it struck the batsman and where it pitched. We don't pretend this is easy, and TV commentators remind us that the umpire does not have the benefit of a slow-motion replay. But there is one case where the umpire frequently decrees 'not out' without appearing to consider the evidence fully.

This occurs when a left-arm bowler is bowling over the wicket to a right-handed batsman (or vice versa, though for some reason umpires seem to be more generous to right-arm bowlers). In this situation, the popular belief is that it is impossible for the bowler to achieve LBW with a ball that goes straight. It is argued that because of the angle from which the bowler releases the ball, it will inevitably miss the stumps if it bounces within the permitted red zone.

It's easy enough to test this with a scale drawing. The width of the wicket is 9 inches and the length of the pitch between the wickets is 22 yards (or 22.86 cm and 20.12 m if you insist on metric units). To make the zones clearer, we have greatly expanded the width of the pitch compared to its length.*

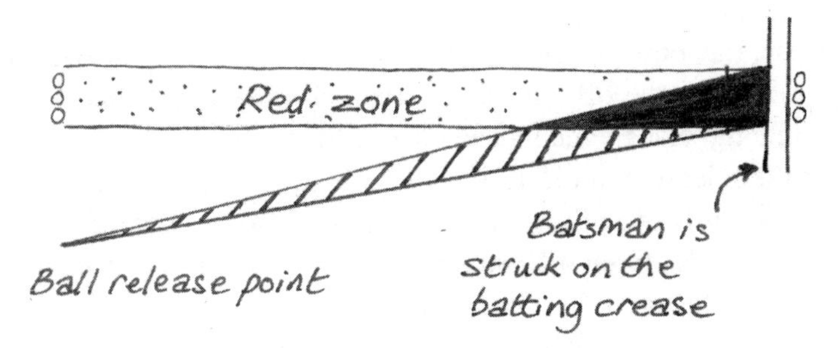

The shaded and black triangles show where the ball with a straight trajectory can pitch if it is going to hit the stumps, so long as it does not bounce too high. If this was a *left-arm* bowler to a *left-handed* batsman, then the ball could bounce anywhere in this shaded area and be legitimate LBW territory.

The area shaded black is where it was hitting the stumps AND pitched in the red zone. This may be small as a proportion of the whole pitch, but

* Note that this doesn't change the relative size of the various zones we are looking at.

it is certainly not zero. Even if the ball strikes the batsman when he is stretching a long way forward (as indicated by the dotted line), there's still quite a chunk of black zone. Furthermore, the closer the bowler gets to the stumps at his end when he releases the ball, the larger the black zone becomes. Bowlers like Alan Davidson or Wasim Akram, or right-arm bowlers like Fred Trueman, who got so close to the stumps that he sometimes knocked the bails off by mistake, were therefore creating far more of an opportunity for an LBW than bowlers like Jeff Thomson who used to sling the ball from a much wider angle.

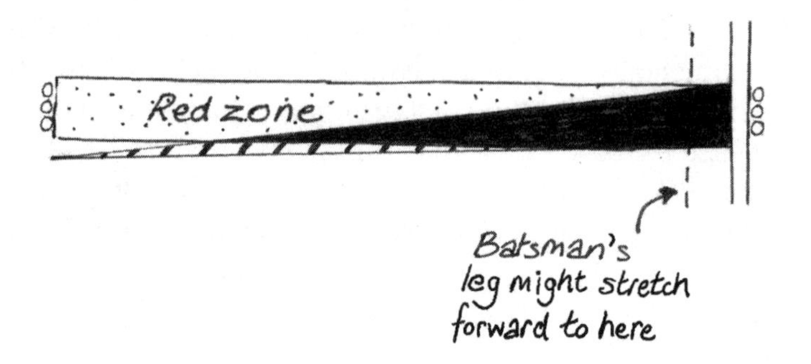

Suppose the bowler releases the ball just nine inches wide of the stumps, aimed at the off stump, and the ball does not deviate sideways. Then it hits the black zone whenever it lands in the batsman's half of the pitch. If the release point is 18 inches wide of the stumps, the black zone begins two-thirds of the way down the pitch – some 22 feet short of the batsman's stumps. Even if he stretches as far forward as he can, he barely improves his chances of avoiding LBW.

In truth, if the ball pitches anywhere short of two-thirds of the way down the pitch, then it is likely to bounce too high to be at risk of hitting the stumps. But that still means that the black zone represents a very significant proportion of all the straight deliveries that would hit the stumps, and that umpires should be more generous to bowlers than they are.*

Perhaps one reason why an umpire hesitates about giving an LBW where the ball may have pitched outside the line of leg stump is down to his own perspective of the pitch. From the point of view of the TV

* The fact that one of the authors is a bowler has nothing to do with this argument.

cameras, which are a long distance away but zoomed in on the pitch, the red zone appears to be a long, thin rectangle. But from the umpire's view, the situation is different. If the red zone were painted on the pitch, then the umpire would see something like this:

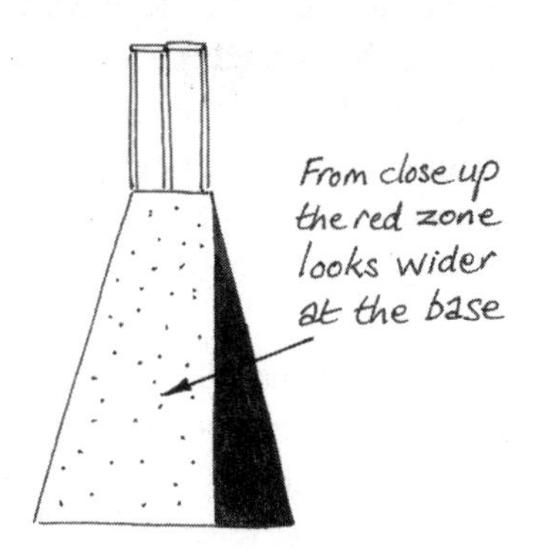

Thanks to perspective, the red zone appears to be narrower near the stumps than it is at the umpire's end. This means that if a ball bounces in the thin black triangle, it might *appear* to have pitched outside the line of the leg stump, whereas in truth it bounced inside. Another reason why umpires give the benefit of the doubt (incorrectly) to the batsman.

Distorted advertising

Rugby and cricket share another example of distorted perspective. The two sports have the dubious honour that their sacred turf is painted with the sponsor's name. Viewed from the blimp soaring above the stadium, the name looks completely distorted, something like this:

But to the TV cameras near pitch level, the word appears in perfect proportion. You can achieve something of this effect by holding the page in front of you and looking up it from a low angle. At one point, the word 'Sponsor' should appear undistorted, fooling the eye into thinking that the word is standing vertically. The designers have compensated for the effects of perspective by stretching the sponsor's name vertically (to compensate for the fact that the cameras are looking at it obliquely) and horizontally (to compensate for the fact that it recedes into the distance). To get the proportions right requires some clever geometry, or a lot of trial and error. The effect can be disconcerting for the TV viewer, since this distorted image creates the impression that there is an ordinary advertising board sitting in the middle of the field. As the player runs to retrieve the ball as it lands on the advertisement, you've half a mind to cry out 'Watch out for that advert!'.

4

IN THE JUDGES' OPINION

The problems of subjective scoring

What do boxing and figure skating have in common? Not a lot, at first glance. But if this ever crops up as a quiz question, then you can arm yourself with at least two plausible answers.

The first is that the notorious Tonya Harding has tried her hand at both sports. Harding, you might recall, became public enemy number one in the USA when she was accused of hiring henchmen to nobble her virtuous skating rival Nancy Kerrigan before the Lillehammer Winter Olympics in 1994. Despite the attempted kneecapping, Kerrigan went on to win silver, while Harding got her comeuppance and ended well out of the medals. A few years later, she made the unlikely move from figure skating to boxing.

The second connection between boxing and figure skating is the source of almost as much controversy. Both sports usually rely on the subjective assessments of a panel of judges in order to rank the contestants.

Rankings of sportsmen are a cause of controversy at the best of times, but when the personal opinions of the judges are permitted to be an important factor, then all hell can break loose. On more than one occasion, the impartiality of the judges has come into question, though the quirky nature of the scoring systems they have had to use certainly hasn't helped. And when it comes to quirky systems, boxing and figure skating have led the way.

The Holyfield–Lewis controversy

Let's start with an example from professional boxing. In March 1999, Evander Holyfield fought Lennox Lewis to determine an undisputed world heavyweight champion. This was good news for a sport that has become so fragmented that at any time there are multiple world champions for each weight, with several boxing 'authorities' each regarding themselves as the official one.

The contest was a classic. By the end of the fight, the verdict of almost all the commentators and most of the crowd was that Lewis had dominated. He'd landed the most punches and had generally been in control in most rounds. However, he hadn't landed a knockout punch, so the fight went the full twelve rounds. The final decision rested with the three judges.

In a huge anticlimax, and to the astonishment of most people who witnessed the fight, the judges announced that the result was a draw. The crowd of over 21,000 booed the result, reckoning it was a fix. But was it? Or were the arcane rules of boxing scoring the main culprits?

In professional boxing, the judges assess the boxers on each round. Each judge keeps their own score all through the fight; they do not know what marks the others are giving. The winner of the round gets 10 points, and the loser receives what are described as points 'in proportion'.

In practice, however, proportion has nothing to do with it. The loser of the round almost always receives 9 points, unless he (or she) has been knocked down and recovered, maybe more than once, in which case the points may be awarded 10–8. But 10–8 is rare, and 10–7 is almost unheard of. Effectively, a boxing match is a series of rounds that are scored 10–9, which might just as well be scored 1–0 to the winner. (If a round is declared drawn, then each fighter gets 10 points, which might just as well be called a 0–0 draw.)

For the Lewis v Holyfield fight, there were three official judges, Eugenia Williams of the USA, Larry O'Connell of Britain and Stanley Christodoulu of South Africa. True to judging form, every round that was not scored as a draw was scored 10–9.

Adding up the points at the end, the US judge made Holyfield the winner by 115–113, the South African awarded the fight to Lewis 116–113, and the British judge scored it 115–115.

If the match had been decided by totalling up all these points, then Lewis would have won by one point. But that's not the way it is done. Each judge merely declares which boxer they found to be the winner, and the majority verdict wins. In this case, one judge declared Holyfield, one declared Lewis, one declared a draw. The overall verdict was therefore a draw.

We've summarised how the judges scored each round in this table. L means the judge put Lewis ahead, H means Holyfield won, and D means a draw.

	Round											
Judge	1	2	3	4	5	6	7	8	9	10	11	12
Williams (USA)	L	L	H	H	H	L	L	H	H	H	H	L
O'Connell (GB)	L	L	H	L	L	H	D	H	H	D	H	L
Christodoulu (SA)	L	L	H	L	L	L	L	H	H	H	D	L

According to Williams, Holyfield beat Lewis by 7 rounds to 5; O'Connell scored them 5–5 with two drawn rounds, and Christodoulu made it 7–4 to Lewis, with one draw. So as with the points, the rounds also went to Lewis by a margin of one.

There are other ways in which boxing might have organised its scoring rules. For example, *each round* might be scored on majority verdict. Under that rule, Lewis would have won the first and second rounds 3–0, and the fourth to seventh rounds by two votes to one, and so on. You can quickly check that this method would have scored the contest as a win to Lewis by 7 rounds to 5.

In fact, just about the only way that these scores could be contrived to represent a 'draw' overall is to use the bizarre system that was actually in place. A fix? No. A flawed scoring system? Absolutely!

Comparing the judges

Part of the problem with the scoring in boxing is that the judges have to make all sorts of subjective judgments, and then combine these into a single result. Credit is given not only for clean punches, but also for 'defence', 'initiative' and 'style'. How do you put numbers against those?

In rounds four and five, the American judge voted for Holyfield while the other two judges both voted for Lewis. This sort of two-to-one decision is bound to happen from time to time, but in this case there is evidence that there was more to it than that.

After the contest, a number of commentators and journalists were asked to give their verdicts on each round. Here is the table of all the verdicts, starting with those of the three official judges, and followed by those of the seven 'media judges':

	Round											
Judge	1	2	3	4	5	6	7	8	9	10	11	12
Williams	L	L	H	H	H	L	L	H	H	H	H	L
O'Connell	L	L	H	L	L	H	D	H	H	D	H	L
Christodoulu	L	L	H	L	L	L	L	H	H	H	D	L

Media Judge 1	L	L	H	L	L	L	L	L	H	H	L	L
Media Judge 2	L	L	H	L	L	L	L	L	H	H	L	L
Media Judge 3	L	L	H	L	L	L	L	L	H	H	L	L
Media Judge 4	L	L	H	L	L	H	L	L	L	H	L	L
Media Judge 5	L	L	H	H	L	L	L	L	L	H	L	L
Media Judge 6	L	L	L	L	L	L	L	L	L	H	L	L
Media Judge 7	L	L	H	D	L	D	L	H	H	H	H	L

Six out of seven media judges scored it as a very comfortable Lewis win, 9 rounds to 3 at least. The other media judge (number 7) scored the match a draw. Were these ten people watching the same fight?

Boxing judges are instructed to score each round on its merits, and not to let their previous decisions sway their opinions in respect of the present round. The unofficial judges seem likely to pay less attention to this factor, and we cannot overlook the idea that they could be in some sort of subconscious collusion ('group think'). Media judges 1, 2 and 3 had identical opinions on every single round!

Even so, the results for two of the rounds are particularly striking. In round five, not one media judge agreed with the official American judge that the round should go to Holyfield. Everyone gave it to Lewis. It looks as though Williams' judgment was at the very least suspect in that round.

Round eight is also interesting. It must have been close, because six judges voted for Lewis and four for Holyfield. But despite this, *nobody* called it a draw. Surely if it was that close, some of the judges should have found it too close to call?

It looks as though the inclination of the judges is always to avoid calling a round drawn if at all possible. Add to this the fact that, except in rare cases, it makes no difference if one boxer wins a round comfortably or by a tiny margin, and it becomes more apparent why boxing can throw up anomalous results.

Amateur boxing, as widely seen in the Olympic Games, does things differently, and better. Judges score each *punch* during four two-minute rounds. If one boxer completely dominates one round, another is even, and he is marginally outscored in two others, he will receive, and deserve, the verdict. If professional boxing followed this lead, then there would be fewer disputes after the final judgment. But maybe that's the whole point. Keep the controversy going, and let the money roll in.

Olympic bias

The more sensible scoring system used in the Olympics removes most of the subjectivity from boxing. However, it's a different story in other Olympic sports, especially those where 'artistic impression' comes into play. The different scoring methods of some of these 'artistic' sports are discussed in the box.

Artistic scoring

It probably comes as no surprise that diving, synchronised swimming, gymnastics and the rest all have different scoring systems. Sports evolve in quirky ways as animals do, and the end results often have no particular logic to them, other than 'that's just the way it turned out'.

For example, in gymnastics, six individual judges mark performances between 0 and 10, in increments of 1/20th of a point, in other words 9.50, 9.55, 9.60, 9.65 and so on. The highest and lowest scores are discarded, and the overall score is presented as the average of the middle four scores, announced to three decimal places. A vaulter whose middle scores are 9.75 from one judge and 9.70 from the other three will end up with a score of 9.737. Strictly speaking, the score should be 9.7375, but the fourth decimal place is ignored (so they are effectively rounding the number down). There are circumstances in which this rounding to three decimal places can alter the final rankings, if two athletes' scores are very close to each other. The chance of this is remote, but could be avoided altogether if the judges forgot the idea of taking the average, and simply added the raw scores together.

In diving, meanwhile, points are scored in much bigger increments of 0.5, again in the range from 0 to 10. This time, seven judges give their verdicts, the highest and lowest scores are again ignored, and the remaining five scores added together. For reasons that baffle the outsider, this total is then multiplied by 0.6 and multiplied again by the 'tariff' of the particular dive, which is a number based on the degree of technical difficulty. In other words:

$$\text{Points} = \text{Judges' Score} \times 0.6 \times \text{Tar}$$

This seems a gratuitously complicated system. The 0.6 is completely irrelevant to the final ranking, and could be omitted. To keep the same level of points as are used at present, the tariffs could just be made proportionally smaller.

Accusations of bias have been rife for many years in everything from floor gymnastics to synchronised swimming. Western countries muttered that the Eastern bloc colluded, the East had a similar belief about the West. Indeed, BBC commentator Alan Weeks got so excited at one event that he reported 'The Russian judge has been suspended for alleged impartiality.' These days, however, if you want to witness blatant collusion, you have to make do with the Eurovision Song Contest (not yet an Olympic sport), where, for example, Greece always gives 12 points to Cyprus, and vice versa.

We noted in the box that when multiple judges are used, the score is sometimes computed after eliminating the highest and lowest scores. This is one clear way in which extreme levels of bias can be eliminated. So, for example, suppose a corrupt judge agrees to mark down a competitor in order to reduce their overall score.

The scores of the seven judges are as follows:

If we take the average of ALL these scores, then it comes to a mediocre 8.771.

But by eliminating the top and bottom scores, the average of the remaining scores is 9.10. Even if the biased judge had offered the outrageous score of 2.5, the final result would have been the same. His plans have been foiled – or at least, their impact has been severely reduced.

However, the biased judge *has* still had an influence. Suppose his 'objective' score had been somewhere in the middle – 9.1, say. If he had voted honestly, the same scoring rules would now give the athlete a score of 9.14, as the score of 8.9 would have been dropped. This honest score is only a tiny amount higher (0.04 points) than the cheating score, but such small margins can make the difference between a gold and a silver, or no medal at all.

Two skating controversies

Let's return to skating, because two of the biggest recent controversies in Olympic judging happened on the ice rink, both of them at the 2002 Olympics in Salt Lake City. And once again, our friends the subjective judge and the dodgy scoring system were at the heart of it all.

After their stunning final performance in the pairs skating, everyone expected the Canadians, Jamie Sale and David Pelletier, to win the gold medal. To the disbelief of the crowd, however, the judges gave it to the Russians, despite a couple of glaring errors from the skaters. This led to booing of the sort that hadn't been heard since, well, Holyfield versus Lewis.

Later the French judge admitted to having been pressured to collude with other judges, and having therefore given her vote to the Russians against her will. Justice of some sort was finally done when the Russians *and* the Canadians were both awarded gold medals. The judge, meanwhile, was fired.

The story was different in the individual skating at the same Olympics, where many felt that justice *wasn't* done. This time, however, it was the scoring system and not the judges that was to blame.

First, here are the bald facts. Figure skating is made up of two contests. The first is the short programme, where the skater has to perform a number of technical moves within a fixed time. The second and more important part is the free programme, where the skaters have more time and much more freedom to design their own routine, in a style of their own choosing.

Here's what happened at Salt Lake City. Michelle Kwan, the darling of America and favourite for gold, was leading after the short programme. Things were still looking good for Kwan as the free skating progressed. In fact, with only one skater left to go, Kwan was still in the gold medal position. As the final skater, Irina Slutskaya, who had finished second in the short programme, stepped onto the ice rink, the positions looked like this:

Gold	Michelle Kwan (USA)
Silver	Sarah Hughes (USA)
Bronze	Sasha Cohen (USA)

Since only one more skater was still to go, you might think that this meant Kwan was at least guaranteed a silver medal. In virtually every other sport, this would be the case. But you underestimate the convoluted figure skating scoring system.

What happened next? As a result of Slutskaya's performance, Kwan dropped from gold to *bronze*, while Hughes leapfrogged above her to take the gold!

Can you figure out what scoring system could possibly lead to this bizarre and counter-intuitive result? If you can, you are bordering on genius. And if you can't, read on – but you might want to arm yourself with a stiff drink first . . .

When you think of the scoring for figure skating, you probably think of all those 5.6s and 5.7s that flash up on the scoreboard. In themselves, however, these scores have never actually been that important. They were merely used as a means of *ranking* the competitors in each programme. This is because, instead of adding together the actual scores for the short programme and the free programme (which you'd think would be the obvious and sensible option), the authorities decided in their wisdom that the *ranks* of the skaters would be added together.

For the system in use at Salt Lake City, the first, second and third place in the free programme would get 1.0, 2.0 and 3.0 points respectively. To reflect its lesser importance, the same positions in the short programme would score 0.5, 1.0 and 1.5 points. The two scores were added together to produce a single number, the winner being the skater with the lowest combined score.

Michelle Kwan won the short programme by a wide margin – in fact all the judges gave her an incredible 5.9 for artistic impression. But all this counted for little. She would have got 0.5 (the winning rank score) even if she'd only beaten the second skater by a tiny margin. The top of the table after the short programme looked like this:

1	Michelle Kwan	0·5
2	Irina Slutskaya	1·0
3	Sasha Cohen	1·5
4	Sarah Hughes	2·0

In the free programme, Hughes did better than Kwan. By the time it got to the final stage with Slutskaya about to skate, the provisional medal table looked like this:

		Short	Free	Total
1	Michelle Kwan	0·5	2·0	2·5
2	Sarah Hughes	2·0	1·0	3·0
3	Sasha Cohen	1·5	3·0	4·5
4	Irina Slutskaya	1·0	?	?

Kwan was still just ahead of Hughes at this stage. If Slutskaya now skated really well, she would win the free programme, and hence the gold medal, and Kwan and Hughes would both move down to silver and bronze. However, if Slutskaya did no better than third then Kwan would get gold. All of this makes sense so far.

There was, though, one way in which *Sarah Hughes* could win gold. It required Slutskaya to come precisely *second* in the free skate. That way, Hughes and Slutskaya would be tied on 3.0 points, Kwan would drop to 3.5 (relegating her to bronze), and Hughes would win gold because in a tie, the one who does better in the free skate gets the verdict. And against the odds, that's exactly what happened. Not surprisingly, many spectators were baffled by the result.

But maybe this story has a happy ending. Prompted by the Salt Lake City controversies, the International Skating Union has set up a new scoring system, under which judges score each element in a programme on how well it is executed. Is there any hope that professional boxing might also reform itself?

5

FASTER, HIGHER, LONGER

The maths of record breaking

Everyone knows that when it comes to athletics, men have an innate physical advantage. That's the main reason why male and female athletics are kept separate. So when in 2004 the press put out a story claiming that women are set to catch up with, and indeed overtake, men in the 100 metres sprint, there must have been more than a few male supremacists choking on their beer.

The story was captioned: 'By 2156, women could be running faster than men', and was based on a light-hearted letter in the scientific journal *Nature*. Scientists had plotted a graph of the Olympic winning times for men and women in the 100 metres sprint, and had concluded that women were closing the gap on men.

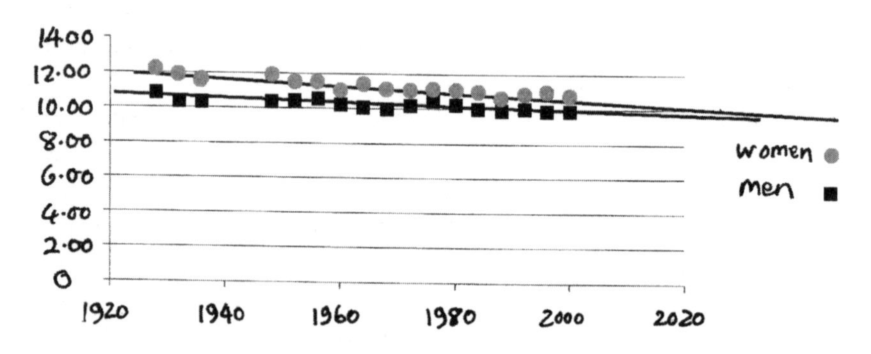

If you look at the graph, there does seem to be a trend. Back in 1928, the winning time for men was 10.8 seconds and for women 12.2 seconds, a gap of 1.4 seconds. By 2000, the fastest man's time was 9.85 seconds, but the fastest woman was only just over a second behind. The time gap has certainly narrowed. But what of the future?

Whatever historic data you have, it is possible to draw a curve through it, and then extrapolate it onwards. The simplest curve of all is a straight line. For the winning Olympic sprint times, the points on the chart do plausibly fall on such a line, albeit with a small amount of random scatter, as you might expect.

However, while it might be reasonable to extrapolate such information a short way into the future, it is an extremely dubious approach for making confident forecasts in the long term, and it's easy enough to show why. Straight lines drawn through the 100 metres data do indeed suggest that within 150 years women will be running as fast as men. But why stop the extrapolation there? Continue the straight line onwards, and the same data suggests that in about 600 years' time, women will be running the 100 metres in *zero* seconds, after which they will begin running in negative time. Unless there are some significant breakthroughs in the field of time travel, this outcome seems a little implausible.

Picking out real trends

So if the 'predictions' published in *Nature* were patently nonsense (and to be fair to the authors, their article was written tongue-in-cheek) then what *is* a reasonable forecast for the fastest times of male and female athletes in the future?

For a start, to make a reasonable forecast you need as much data as possible. The 100 metres forecast was made on just seventeen data points, drawn from the performances of individual athletes. If you want a trend, then a much safer statistic is to look not at the *extreme* performances each year, but the *average* performances. They are less susceptible to freak events that mislead. One way to do it is to plot the average of the ten fastest 100 metres times each year. When this is done, the answers are revealing.

The graph of men's sprint times still follows what looks like a linear decline, particularly after the Second World War, though it appears to flatten out very slightly in the later years. For the women, though, the shape of the graph takes a much more dramatic turn. Sometime around 1984, the line suddenly appears to level off. If anything, the gap between men's and women's performances increases after that year.

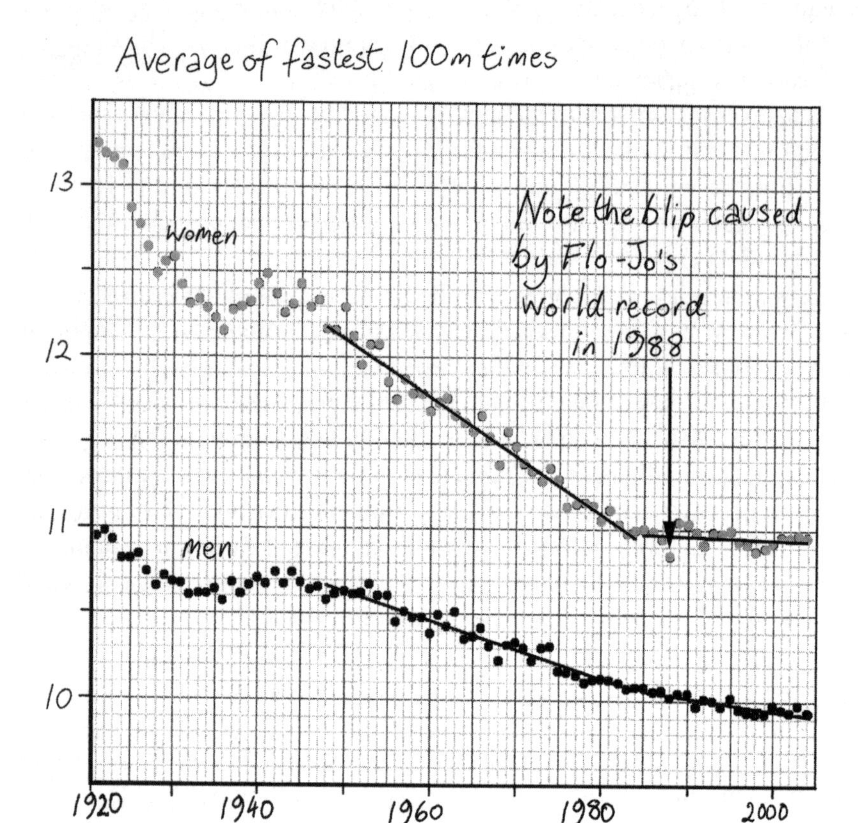

What could the explanation be for the kink in the 1984 graph? Maybe the answer is drugs. During the 1980s the level and effectiveness of drug testing increased dramatically. Eastern European athletes who had been brought up on steroids were suddenly excluded. Apart from a spectacular run by Florence Griffith-Joyner ('Flo-Jo') of 10.49 seconds in 1988, the average sprint time for women has practically remained the same for over 20 years.

The number of women athletes has risen more dramatically than the number of men over the last century, and the more athletes there are, the more exceptional performers there will be to choose from. This increase in the number of athletes probably explains the extremely steep improvement in women's sprinting before the Second World War. But the number of athletes outside China has now reached a plateau, partly because the population in the West is no longer growing. And in the next century the forecasts suggest that the number of people in the athletics age range of 18–30 will dramatically decline *worldwide*. Common sense suggests that the smaller the pool of athletes, the fewer potential record-breakers will emerge.*

When is a record not a record?

There are two types of world record: those that should be allowed, and those that shouldn't. Unfortunately, deciding which records belong in which category can be an inexact science.

For example, in 1996 Obadele Thompson ran the 100 metres in an astonishing 9.69 seconds. Nobody had ever run anything like as fast before, even with the aid of drugs. So why did it not count as the world record? Because, by all accounts, the wind blowing down the track that day was approaching gale force, and wind-assisted runs don't count. Or at least, they don't count if the speed of the wind blowing down the final straight is greater than 2 metres per second (2 m/sec).

The official world record at the start of 2005 stood at 9.78 seconds to Tim Montgomery, achieved with a following wind smack on the legal limit. Maurice Greene, whose time of 9.79 seconds fell to Montgomery's run, might well have had a rueful smile – he set his record with only a hint of help, a barely perceptible light breeze of 0.1 m/sec. Indeed, Greene also recorded a time of 9.82 seconds, *against* a headwind of 0.2 m/sec. Which performance, adjusted for wind speeds, was the best?

A following wind reduces the amount of wind resistance that a sprinter feels. (Even with a strong tailwind, a sprinter running at top speed will feel wind in his face, not on his back.) As a rule of thumb, for each metre per second of wind speed, a sprinter will trim 0.06 seconds off their time for 100 metres flat.

On this basis, Montgomery's world record translates to an equivalent time of about 9.90 seconds if run in dead calm, while Greene's 9.79 (with rounding) gets adjusted to 9.80 seconds, and his time when running into the headwind comes down from 9.82 to 9.81 seconds. So the *wind-adjusted world record* for 100 metres is hereby declared as 9.80 seconds – and congratulations to Maurice Greene.

* Although physiology suggests women are unlikely to overtake men in the short events, the situation may be different in the very long events because female endurance is typically superior. If you wish to survive for several freezing nights on the North Face of the Eiger, take the precaution of being female.

There have also been question marks over some women's sprint records, not least the late Florence Griffith-Joyner's world record for the 100 metres of 10.49 seconds. Experts have questioned this remarkable time for a number of reasons, including a suspect wind-gauge reading of zero. And there is indirect evidence to support those doubters. Flo-Jo's second-best (legal) times were 10.61 and 10.62 seconds, both with following winds. She also ran a time of 10.54 seconds, but with a following wind of 3.0 m/sec, and so that run turns out to be worth around 10.71 seconds without the wind. Her world record, if really achieved in complete calm, was some *two-tenths of a second better than every other run of her life*. Hmm.

Incidentally, world records are quoted to two decimal places, even though the sophisticated timing equipment can measure much more accurately than that. The time recorded is when the trunk of the athlete, not just any part of the head or body, crosses the finish line. At the speeds of Olympic sprinters, a difference of 1/1,000 of a second represents less than one centimetre. Whether the athlete happens to be breathing in or out at the moment of finishing could become the decisive factor! Awarding a dead heat, rather than relying on a blurred photograph, is not a cop-out.

Mother Nature can help athletes to break records with more than just a following wind. In the 110 metres hurdles in 1993, Colin Jackson set a world record of 12.91 seconds, helped by a small, but legitimate following wind. Astonishingly though, on another occasion Jackson also achieved the time of 12.97 seconds *against* a strong headwind of 1.6 m/sec, which would have smashed his world record if we make an adjustment for wind speed. There is, however, a catch.

Jackson set this record at Sestriere, some 2,000 metres high in the Italian Alps. At such altitudes, the air is thinner, its resistance is less, and this can make a significant difference to a sprinter's time. There is a strong case for establishing two categories of world record, those achieved at around sea level, and those at high altitude. Indeed some athletes have resolutely declined to seek opportunities to break existing records by deliberately not competing at high altitude.

How to throw further

There is another small benefit of high altitude, aside from thinner air. The further you are from the centre of the earth, the lower the force of gravity. And there are some record-seekers who can benefit from this even more than the runners do – like shot putters, for example.

If a shot putter wants to break the world record, what tactics should he use? As a starting point, he could do worse than copy tactics that have been known to naval officers since the time of Francis Drake. The aim of a warship captain was to wipe out the enemy ships without taking any hits himself. He could do this if he could fire his cannon balls further than the enemy, and so there was a lot of interest in the mathematics of increasing the range.

What, then, is the optimal angle to point a cannon if you want to fire a ball as far as possible? Intuition helps to point to the best solution. If the cannon crew aimed the gun dead level, then the ball would begin veering down towards the ocean as soon as it left the barrel, under the influence of gravity. If they aimed vertically upwards, then their own sailors would be leaping for cover. Clearly the 'optimum' angle to point the cannon is somewhere between the two. As it happens, the maximum theoretical range when aiming at a horizontal target, ground level to ground level, is to fire your cannon at 45 degrees (so long as you ignore wind resistance).

All of this is similar to the situation faced by the shot putter. Even the projectile is the same – the original 'shots' that were being 'put' by athletes were cannon balls. So the 45 degree rule is not a bad guide for a shot putter learning his trade, though there are other ways to improve the range, too.

Common sense will tell you that if you want to project the shot as far as possible, it helps to heave it as fast as possible, from as high as possible, and with as little gravity as possible. The maths shows that these factors are not equally important, however. A small increase in the speed of the ball has a much bigger impact than a small increase in the height of the thrower. Increasing the speed of release by 1 per cent will add 2 per cent to the result; release the shot from 1 per cent higher, and the distance goes up by only 0.1 per cent. It is more important to be strong than to be tall to succeed in this event.

The strength of gravity also influences the range of a shot. Lower gravity can be found either by travelling towards the equator (where the earth bulges out) or by climbing to altitude. This gives record breakers *two* reasons to prefer Mexico City over sea-level Helsinki. The pull of gravity is about 1 per cent less – worth about 20 cm to a top-class shot putter.

There is, in fact, a formula for the horizontal distance a cannon ball will travel. Heeding the advice of Stephen Hawking that every equation halves the readership of a book, we won't show the formula here, but if you like that sort of thing, you'll find it in the Appendix.

There are some important differences between putting a shot and firing a cannon (on top of the obvious ones to do with noise, life expectancy and so forth):

- a ship's cannon aims at a target that is the same height above sea level as itself, whereas a shot putter releases the ball about two metres higher than the point where it ends up;
- unlike a cannon, which is equally efficient whichever direction you point it, the human body finds it easier to send the shot horizontally rather than vertically, so a higher speed is possible at lower release angles.

For both of these reasons, a shot putter can increase the range of his throw by releasing the shot at a slightly flatter angle than the 45 degrees that is best for a cannon. In fact the next world record may well be achieved by a shot putter who releases the ball at 42 degrees to the horizontal. Remember, you read it here first.

Jumping higher

Sometimes it takes radical thinking to break a record. The best recent example is surely Dick Fosbury, who at the 1968 Olympics changed the face of high jumping overnight. He won the gold medal using the technique now known as the Fosbury Flop, and it is rare to see any other style of high jumping in international athletics today.

The secret behind the Flop is that it requires less energy to clear the bar than a traditional straddle-style jump. In early high-jump competitions, the popular style of jumping was known as the scissors, because of the motion of the legs:

The problem with this style of jumping was that to clear the bar, the jumper had to lift his centre of gravity considerably higher than the bar.

The rules later changed to allow jumpers to go head first, helped by the introduction of a soft cushion on the other side of the bar. This led to new techniques including the Western roll and the straddle, both of which enabled the jumper to hug much closer to the bar. In fact, in the case of the straddle, it was just about possible to dangle the legs and arms over the bar in such a way that the jumper's centre of gravity *never* had to go higher than the bar. It could be argued that the jumpers weren't going higher at all, they were simply being smarter:

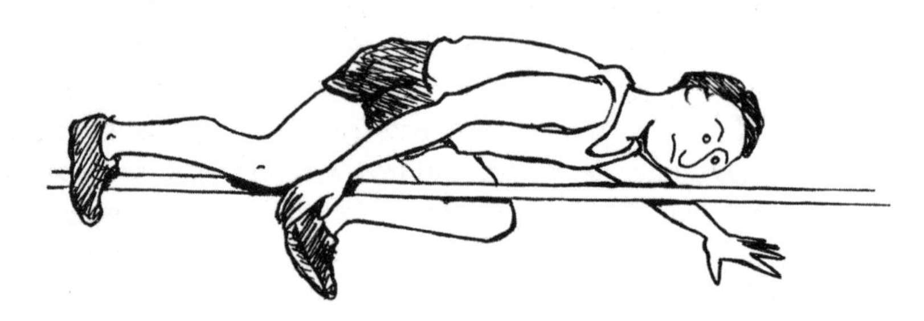

All of this culminated in the Fosbury style, where the athlete twists on take-off to jump with his back to the bar, arching his back, with his feet the last points to pass over the top. The more he bends his back, the lower his centre of gravity becomes.

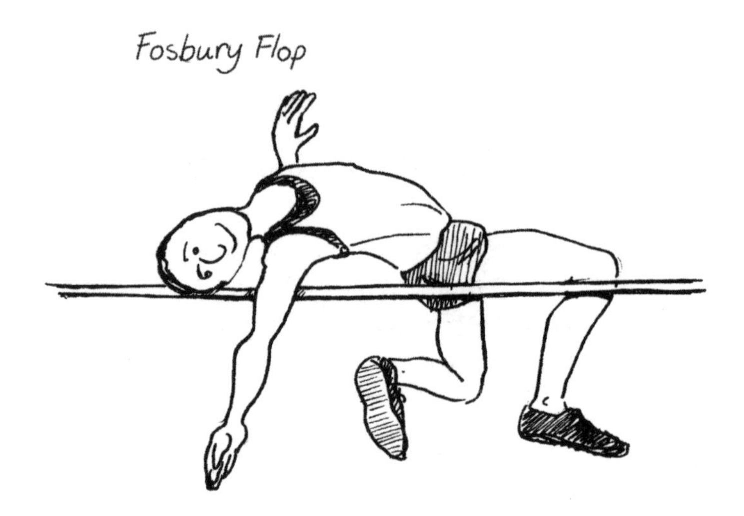

Tall people have an advantage with the high jump because their centre of gravity is further off the ground to begin with, so perhaps the real measure of quality is the difference between the height cleared and the athlete's own height. Under this measure, Franklin Jacobs, only 1.73 m tall, takes the honours, having cleared 2.32 m for a 59 cm differential. Javier Sotomayor's world record is 2.45 m, just 52 cm more than his own height. (The differences in body shapes of men and women are such that the corresponding height differential the best female high jumpers attain is about 30 cm.)

Although the Fosbury Flop has become the industry standard for high jumpers, there are those who question whether it really is superior to the straddle. Both jumps can keep the centre of gravity low, but the athletes still need to be able to lift themselves, and the key to doing this is in the speed of approach and the amount of thrust. It has been argued that a more powerful spring can be achieved running forwards than by twisting around for the Fosbury, and this extra power compensates for the extra lift that is needed to do the straddle. Indeed, two mathematicians have asserted that the next world record will be achieved with a straddle and not a Fosbury.

A decline in record breaking?

However, the next world record may not come for some time. At the start of the chapter we looked at how the trends appeared to show a constant level of improvement. But are we really going faster, longer and higher?

There is evidence that in some sports, the trends may be pointing the other way, with the high jump being the most dramatic.

On page 54 is the graph showing the average height of the best ten jumps each year in the high jump.

As with the 100 metres graphs earlier, the high-jump graphs kink markedly in the mid 1980s. In this case, the men's graph begins to point downwards. If you chose 1984 as a starting point, and used a straight line to forecast the future, you could make a case for men jumping lower than women by the year 2156 – not so much because the women were getting better but because the men were getting markedly worse.

And why stop there? According to this forecast, in the year AD 3339 men won't be able to get off the ground at all. If you are selective enough with your data, and prepared to abuse the central principles of statistics, you can extrapolate to reach whatever conclusions you want. But don't expect anybody to take you seriously.

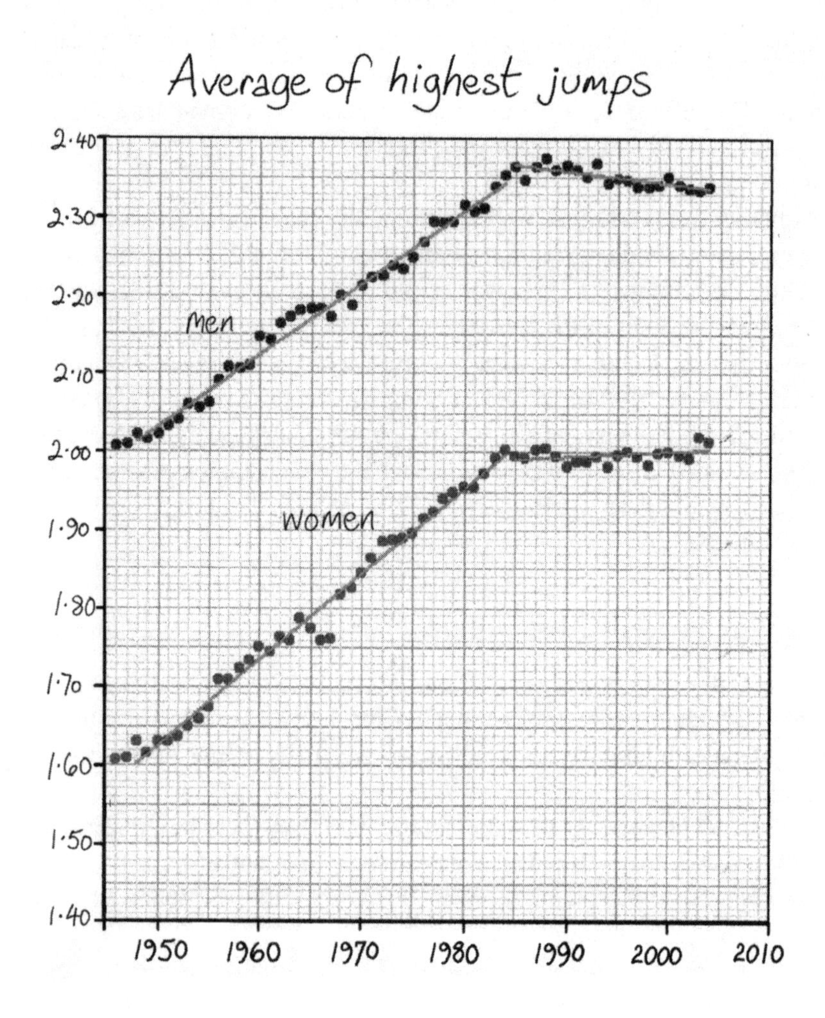

6

THE TOSS OF A COIN

How calling heads can influence a game

On Saturday 11 May 2002, Stoke City and Brentford lined up to play the key match of their season. This match, in front of a crowd of 30,000 at the Millennium Stadium in Cardiff, would decide which of them would be promoted to the First Division.

But there were some who believed that the result was a foregone conclusion, no matter how the two teams played. Earlier in the week the teams had tossed a coin to determine which changing rooms they would occupy on the day of the final. Brentford had won the toss and elected to get changed in the stadium's North Dressing Room.

In all eleven previous finals and play-offs held in the stadium, the team in the North Dressing Room had gone on to win the match. The South Dressing Room seemed to have developed a curse, which had brought down some of the best teams, including Arsenal and Chelsea. By now everyone in football knew that the north room was the lucky one. No wonder the Brentford players were relieved when they won the toss, because now history decreed that they were destined to win the match.

Unfortunately for Brentford, Stoke City didn't read the script. Stoke won the match 2–0.

What had brought about this remarkable change of fortune that had turned the South Dressing Room from unlucky to lucky? Some people attributed it to the work of a feng shui expert who had been called in a couple of months earlier. He had identified 'negative energy' coming from the TV interview studio that backed onto the changing room, and on his recommendation a striking mural had been painted on the South Dressing Room wall to counteract it. Clearly Stoke had won thanks to the new mural. Those who argued this conveniently forgot that the feng shui expert's consultation happened in March. The lucky mural was no help to Cheltenham Town, for whom the North Dressing Room hoodoo continued until 6 May, a few days before Stoke's victory.

Others had a simpler explanation. Stoke won the match because they scored two goals and Brentford scored none. But where's the romance in an explanation like that?

Streaks

'Streaks', lucky or otherwise, are the meat and drink of sports reports. Look in any newspaper on a Saturday morning, and the chances are that at least one of the preview articles will refer to an unbroken sequence that has occurred in previous encounters between the competitors. Sometimes there might well be significance in this. For example, if Lleyton Hewitt has beaten Tim Henman every time they have played a tennis match, it's probably a sign that Hewitt is the better player. Even if he isn't, Hewitt may now have a psychological advantage over his opponent simply because the pressure is off him to prove he can win these matches.

Whether such streaks are significant or not, there is always a carefully selected starting point for the sequence. If you read that Michael Owen has scored in his last five games, you can be quite certain he did *not* score in the match six games ago. And keep a look out for the weasel words: an emphasis that Arsenal had a run of 49 *League* games without losing should tell you that they did lose some Cup games during that period – for otherwise we would read about their (longer) undefeated run of League and Cup games. Reporters try hard to convince readers that something very unusual is happening.

When these streaks are mentioned, the implication is that they will have a bearing on the next encounter, but this is often not the case. Lucky streaks are bound to happen, in just the same way that if you toss a coin, there will be periods when it comes up heads several times in a row.

There is a common fallacy with tossing coins, namely that a long string of heads might have some bearing on what happens with the next toss. The truth, however, is that each toss of a coin is entirely independent of what has gone before. Coins have no memory of their past performance, there is no sense in the notion that an excess of heads one day makes tails more likely on the next day. If you get a sequence of heads, then it is true that eventually your luck will change, but exactly *when* it will change is another matter.

This has a direct parallel with the story of the North Dressing Room at the Millennium Stadium. In Cardiff, some suggested that because the last eleven tosses of the coin (for which you should read 'the last eleven winners of the match') had been 'heads' (the North Dressing Room), then the next one would be too. Others voiced the opposite and equally fallacious argument that because the North Dressing Room had won so often, a change of luck was 'due'.

The chance of a long streak

Let's suppose, as is reasonable, that any connection between choosing the North Dressing Room and the result of the match is purely down to chance. What would be the probability of the North team winning eleven times in a row?

The chance of winning one match would be ½. The chance of winning two or more in a row can be seen by presenting the possible results as a branching tree.

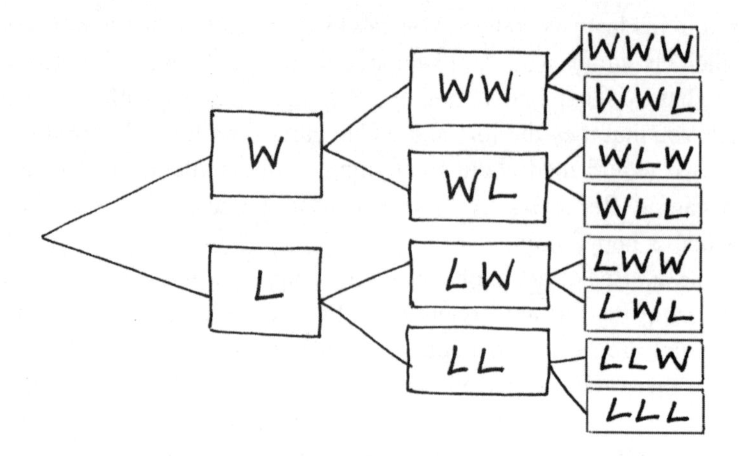

The chance of winning two in a row is 1 in 4, which is the same as ½ x ½, while the chance of three successive wins is 1 in 8, or ½ x ½ x ½.

You can extend this method to calculate the chance of winning the toss eleven times in a row, which is:

½ x ½ x ½ x ½ x ½ x ½ x ½ x ½ x ½ x ½ x ½

more concisely expressed as $(½)^{11}$ or 1 in 2,048. That is seriously unlikely. No wonder the football world was spooked.

This raises the question of *how likely* it is that a long lucky (or unlucky) streak of coin tosses will occur at some stage. Before we answer this question, have a go at some detective work by trying to spot the fake in the box opposite.

We will answer this little quiz in a moment. But first, how do you go about identifying a genuinely 'random' sequence of coin tosses? One way is to look at the longest streaks of heads and tails (or wins and losses). Remarkably, there is a formula that is a very good approximation for how long you might expect the longest streak to be:

If you toss a coin **N** times, then expect the longest streak of heads or tails to be a bit more than the logarithm of **N** in base 2

Spot the fake

Below are four sequences of 32 tosses of a coin. L indicates that the caller lost the toss and W indicates he won. Now here's the catch. Two of the sequences are genuine – in fact they are a sequence from a real sporting contest over a span of 32 years. The other two are fakes – two people attempting to come up with sequences that look genuine. Obviously, all four sequences are possible; but now you know that two of them are fakes, which pair do you assess as more likely to be these fakes?

(1) L W L L L L L L W L W L W L L L W W L L W L W W L L W W W W W W

(2) W L W L W W L L L W L L W W W W L W W L W L W L L L W W L L W L

(3) W L L W W L L L W L W L W W L W L L W L L W W W L L W L W L W W

(4) L W L W L L W L W W L W L L W L W W L L W L L W L W W L W L L W

If you want to know more about the reasoning behind this formula, or need a reminder of what a logarithm is, see the Appendix.

Now back to the longest run of heads or tails. To estimate this, you take the logarithm (base 2) of the number of times you toss the coin and round it up to the nearest whole number. The longest streak could be more or less, but this is a rough average. So if you were to toss a coin 16 times, then (according to the formula) you might reasonably expect there to be a run of at least FOUR (= $\log_2(16)$) or more heads or tails somewhere in the sequence. If you tossed a coin 32 times, then look for a run of at least FIVE heads to occur, and if you tossed a coin 16,384 times, then don't be at all surprised if a run of FOURTEEN consecutive heads occurs at some point.

This doesn't mean that these long sequences are guaranteed to happen. After all, it's possible for the sequence to go heads, tails, heads, tails, heads, . . . indefinitely. However, this is just as unlikely as every toss coming down heads. Taking everything into account, if you toss a coin 32 times, the most likely outcome is that the longest streak of heads or tails will be either five or six.

Now look back at the genuine and fake sequences. In the first two sequences, the longest streaks of W or L are six and four respectively. These are both close to $\log_2(32) = 5$. And as it happens these are the genuine sequences. In fact they cover the periods 1973–2004 and 1920–1957 respectively, and indicate whether or not Cambridge won the toss in the Boat Race. Neither of the other two sequences has a streak longer than three. These last two are the fakes.

To many people the genuine sequences look like the fakes. It seems that people who try to create 'random' sequences switch from W to L or vice versa if they feel that one of them is happening too often. But reality is different, as the Boat Race figures illustrate. And the maths backs it up.

Hussain's losing streak

If you think the sequence of eleven losing finalists at the Millennium Stadium was impressive, consider the case of Nasser Hussain, the former captain of the England cricket team. Over a period during 2000–2001, Hussain captained England in fourteen international matches. He lost the toss in every single one.

Early in the streak, the press picked up on the fact that Hussain was losing more than his fair share of tosses, which made each subsequent loss of the toss all the more newsworthy. To add to the drama, Hussain was injured during the Ashes series of 2001, after he'd lost seven tosses in a row. Michael Atherton took over as captain – and promptly won the toss. When Hussain returned to the side he proceeded to lose the toss another seven times. Incidentally, Hussain insisted on calling heads each time he was given the choice. It might seem like this 'stubbornness' in sticking with the same call was part of the problem, but of course his chance of winning the toss would have been no different if he had alternated between heads and tails each time.

The chance of losing the toss 14 times in a row is 1 in 2^{14}, a number we calculated earlier as 16,384. These odds are so remote that it is natural to begin to look for some cause other than chance that might have brought about this streak of misfortune. But it is important to look at the bigger picture. Hussain captained England 101 times in total. According to our calculations, if somebody tosses a coin this many times then there is roughly a 1 in 180 chance that they will encounter a streak of 14 or more wrong (or right) calls in a row at some stage. It still makes Hussain's run remarkable, but 1 in 180 is very different from 1 in 16,384. Moreover, Hussain was just one player who had captained his country many times. Maybe it isn't all that surprising that, among all these captains, one of them should seem to be desperately unlucky.

Does winning the toss matter?

Most sporting contests begin with a toss of a coin or drawing of lots. For games between two individuals or teams, the coin determines the choice of ends, which side goes first, who starts from the 'favoured position', who gets to wear their first choice of shirt, and so on. Which all begs the question: to what extent does winning the toss *matter*? If the toss is significant, then all the post-match garlands for the winners become tainted by the fact that it may not have been sporting skill but rather pure luck that brought them victory.

Often, the advantage of winning the toss will be negligible. In football, for example, the captain who wins the toss usually decides on which direction his team want to kick in the first half. Teams usually prefer to be kicking towards their own supporters at the end of the game. There may be some psychological benefits from being roared on in the closing stages, but since the opponents' fans will be at the opposite end of the ground both teams will therefore enjoy a similar boost at the same stage of the game. Common sense says that sports where there is complete symmetry in the conditions will offer minimal advantage to the side that wins the toss.

But even in games of two halves like football, hockey and rugby, there can be a degree of asymmetry, in particular due to the weather conditions. One common factor that can influence a match is wind. This can have a noticeable bearing in rugby because it determines whether a kick from a particular distance will be comfortably within range or almost impossible. Both teams will play half the match with the wind and half the match into it.

Would you prefer to play with the wind in the first half or the second? There is no sure-fire mathematical answer, but you can balance different factors. In favour of playing with the wind first, there is the psychological advantage of taking a lead in the match; the fact that the wind may drop later; and any belief that this referee is inclined to give lots of penalties early in the match (to remind the forwards that some of the rules are meant to be obeyed). In favour of the alternative, more points tend to be scored later in the match, as players tire – so if the wind advantage is worth 60 per cent of the points, you get 60 per cent of a larger number; and usually, more 'stoppage' time is added on in the second half than the first. Our (pessimistic?) inclination is to take the wind first – a bird in the hand . . .

In one particular case, the asymmetrical conditions irrefutably favour the team winning the toss. On an English winter's afternoon, where the pitch points east/west, the team that plays towards the west in the first half will have an overall advantage. For as the match progresses, the sun will approach the horizon, and will increasingly be in the eyes of the west-playing team as they try to concentrate on the game. High punts towards the rugby full-back, and long-distance shots in soccer, are more likely to be rewarded if the opponent has the sun in their eyes.

The situation becomes a lot more complicated when the advantage clearly alternates between the opponents, for example when the two sides take it in turns to serve. In tennis, for example, the winner of the toss has the right to serve first and will usually do so because the server normally expects to win a service game (at least in professional tennis – the same is not necessarily true of the tennis on display in the local park). However, this advantage doesn't last, because of the rule in tennis that there has to be a clear margin of two games before winning a set. If the game gets to 6–6, then the tie-break is also designed in a way that doesn't mathematically favour either player: the first to serve only gets one point, followed by the opponent having two serves. Neither player can therefore win the tie-break unless she wins more points on the opponent's serve than she loses on her own.

There are, however, examples in sport where the winner of the toss *does* have more opportunities to score than the opponent. In squash and badminton, the rules outside the USA have traditionally been that only the server can score a point. Therefore, the player who wins the toss gets the chance to score a point immediately, whereas if the non-server wins the rally, he merely wins the serve and therefore the opportunity to win the point after that. This gives a small, but measurable advantage to the person who wins the toss. Take two players of equal ability: the scoring system means that the player who serves first will actually *score* the first point two-thirds of the time. This advantage is eroded as the set progresses and service changes hands, but our calculations suggest that the player who wins the toss in squash should win the set around 53 per cent of the time. Unlike tennis, the server in squash has no significant advantage in *winning* a point, so a fairer way to start a game of squash using the old rules would be for the first point to be used simply to decide who will get to serve the second point.

Even more of an advantage can apply in top-class darts, where the player who throws first is considerably more likely to win a game than his opponent. We say more about this in Chapter 7.

The toss and cricket

Which sport is most influenced by the winning of the toss? Many believe that it is cricket.

The captain who wins the toss in cricket decides whether his team will bat first or bowl first. Usually the pitch is in its best condition on the first day. As the match progresses, the bowlers' spikes increasingly scuff up

the ground while the sun steadily dries out the pitch, causing cracks to widen and the soil to crumble. As the condition of the surface deteriorates, the behaviour of the ball as it bounces on the pitch becomes less predictable, which means it gets harder to bat. For this reason, the captain who wins the toss usually chooses to bat first.

The perceived importance of the toss is confirmed by the fact that in cricket, the written reports will almost always record who won the toss. It is far easier to look up who won the toss in an international cricket match than in any other sport. When we tried out the innocent phrase 'won the toss sport' with Professor Google, over ninety of the top hundred hits were about cricket matches.

But in limited-overs cricket where the weather doesn't play a part, the importance of winning the toss is often exaggerated. A study of over 400 one-day international matches in the 1990s found that each of the nine countries tended to win the same proportion of games whether they won the toss or lost it. In other words, winning the toss had almost no bearing on the result. This doesn't rule out the possibility that weather conditions can seriously favour one team, particularly in day/night cricket when evening dew can sometimes make the pitch very lively, but the proportion of games where this affects the result appears to be very small.

Another study of 151 Test matches from 1997 to 2001 concluded that winning the toss was of no real help in five-day matches either – the team winning the toss did marginally worse than its opponents! Have we all been wrong about the importance of winning the toss in cricket for the last hundred years?

Heads on the river

It seems that the toss in cricket isn't as significant as folklore would suggest. And except for situations where particular weather conditions are at play, its impact in other sports is still only marginal. Perhaps one sport where there is a distinct advantage in winning the toss is the Oxford v Cambridge Boat Race.

The toss allows the captain to choose which side of the river he wants to row on, the Surrey side or the Middlesex side. Because the River Thames has significant bends in it, the boat on the Middlesex side gains the early advantage because it is on the inside of the curve. But unless the Middlesex boat gains a lead of more than a length and is able to move to

the middle of the river, the Surrey boat more than regains that advantage when the river bends the other way.

Captains usually opt for the Surrey side, and on the face of it this seems a good decision. Captains who chose Surrey have won nearly 60 per cent of their races. Captains who chose the Middlesex side have only won 50 per cent of their races. Does this mean the winner of the toss should always choose Surrey? Not necessarily. A team that feels it is the underdog may think its only chance is to steal a lead on the first bend. A 50–50 chance of winning by choosing Middlesex is better than what might only be a 30–70 chance if that particular team were to choose Surrey. So a Boat Race captain needs an element of skill and judgment when deciding what to do after he wins the toss.

In the 80 races from 1920 onwards, the winner of the toss won the race 44 times, a 55 per cent success rate. This sample is too small for any significance to be read into the result, but if after another 200 races the 55 per cent success ratio is maintained, we will be confident that the annual Boat Race is one event where winning the toss really does matter.

7

ONE HUNDRED AND EIGHTY!

Where to aim on the dartboard

Is darts a sport? It's a question that vexes many people, particularly because the most common venue is a pub, not somewhere that is immediately associated with people at the peak of their mental and physical fitness.

Nonetheless, darts has rival world championships, it features in encyclopaedias of major sports, and many hours are devoted to it on TV sports programmes. That's good enough for us. (Also, in 2005, Sport England – formerly the English Sports Council – officially recognised darts as a sport, but this is unlikely to settle the argument.)

As well as requiring a high level of composure and hand–eye co-ordination (like archery and shooting), darts has the extra dimension of demanding a high standard of mental arithmetic. Perhaps only limited-overs cricket approaches darts for the mental calculations needed while the game is in progress.

In the classic form of darts, each player starts with 501 points, and the objective is to reduce this to exactly zero, throwing three darts in each turn and deducting these scores from the target. The first player to reduce their score to zero wins the leg, though there is the added constraint that the final dart has to be a double (which includes the bullseye, as 'double 25'). The mental calculations are needed not only to do subtractions after each round, but also to work out an efficient way of finishing the game.

The normal strategy at the start of the game is to aim for the biggest score available, treble 20. The professionals land most of their darts in the '20' segment, and so their calculations are simple. Lesser players, however, are more erratic, and so need to get used to computing sums like 417 minus the sum of treble 5, 18, and 11. It is ironic that for centuries, schoolmasters prescribed difficult sums as punishment for pupils who weren't performing well, yet in bars and pubs across the world, millions of folk happily volunteer for this punishment every evening.

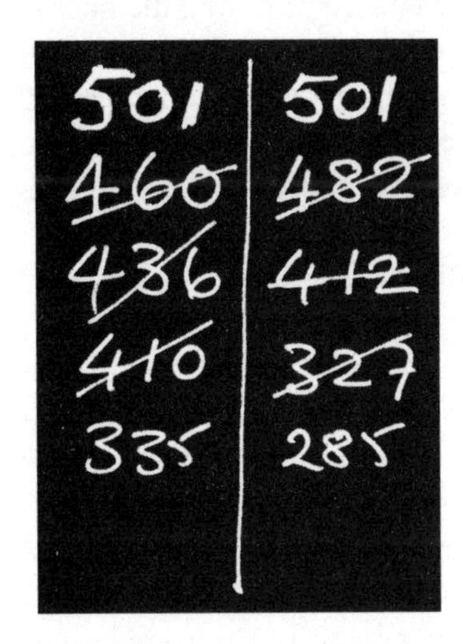

Tactics for pub players

Pub darts players aren't only gluttons for arithmetical punishment, they also suffer from a degree of self-delusion in trying to copy the tactics of the professionals.

Top players are very consistent, and aiming at the centre of the treble 20 is indeed their best strategy. But pub players are far more variable, and just missing the 20 segment will result in low scores of 1 or 5. So while pride or vanity induces everyone to copy the tactics of the professionals, in truth most players would be better off adapting their tactics according to how *consistent* they are. The more variable the accuracy of a player's throw, the less sensible aiming at treble 20 becomes, because the average score for a dart will be dragged down by the low-scoring neighbouring segments.

For a player who would expect to land a dart in the right segment nearly two-thirds of the time, and one of the neighbouring segments the rest of the time, aiming at *treble 19* is likely to give the best average score. But as accuracy reduces to hitting the segment half the time (which is quite possible after a beer or three), *treble 14* becomes increasingly attractive as the basic strategy. Even if you miss the 14 by two segments, a dart won't do worse than 8 points. Fairly erratic players should aim at *treble 16*, because that quarter of the board has the best overall average. And if your darts are truly appalling, you should simply aim at the bull. This does at least maximise your chance of hitting the board.

In fact for many darts players, there is a sneaking suspicion that because of their inaccuracy, actually trying to aim the dart is not a benefit at all, it is a hindrance. One way to check this out is to calculate what score you would make if you just blurred your eyes and aimed randomly at the board.

To make the question meaningful, we have to be a bit more precise because, if darts really are thrown completely at random, many of them will miss the board. So, we will assume that the three darts all land in the scoring part of the board, but the chance of falling in any particular segment is proportional to the area of that segment. And, for simplicity, we ignore the thickness of the wire that separates the segments.

The double ring occupies about 9 per cent of the whole scoring area. But *if* a randomly thrown dart lands in a double, according to our assumption it is equally likely to hit any one of the twenty segments, which score 2, 4, 6, . . . 40 respectively, i.e. the *average* score for a double is 21.

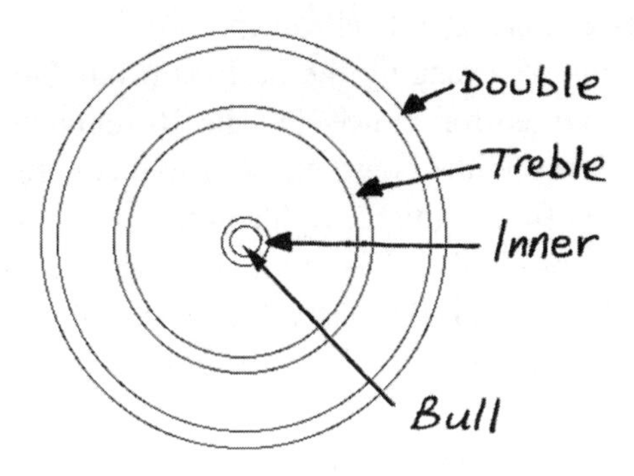

Repeat this calculation for each distinct region of the board. The treble ring occupies 6 per cent of the scoring area, and the average score for a treble dart is 31.5. For singles, the average score of one dart is 10.5, and their segments take up 84 per cent of the scoring area. Finally, the bull (scoring 50) is a mere 0.14 per cent of the area, while the inner (scoring 25) occupies 0.74 per cent.

To obtain the average for just one dart, we weight these average scores according to the frequencies at which they occur. The result is an average score of 12.8 points per dart. Multiply by three for the average over three random darts, and the answer is about 38.5 – call it 40, since this is such a rough and ready calculation.

So, 40 points is the touchstone. As long as you average higher than 40 points in a single round, you are doing better than a random thrower. But if you average much under 40, keep it to yourself: a machine that scattered its darts randomly on the board would score higher than you.

Anyone for blindfold darts? (That finishing double may take a little time . . .)

The unforgiving numbering scheme

Part of what catches out run-of-the-mill players is the clever way in which the numbering on the board has been designed. The numbers are placed in such a way that the biggest scores are always adjacent to low scores, which act as a trap for the overambitious, a bit like the bunkers on the Old Course at St Andrews.

The invention of the dartboard numbering is credited to a carpenter from Bury called Brian Gamlin in about 1896, and no one has seriously proposed any change to it since. But if you want to separate out the class players from the rest, is it the 'best' order?

In terms of ways of arranging the dartboard numbers, you could say that Gamlin was spoiled for choice. Put the 20 segment in its traditional position, and work round the board clockwise: there are nineteen choices for what goes in the first segment, eighteen for the second, and so on. So the total number of possibilities for arranging the board is 19 x 18 x 17 x . . . x 2 x 1, which is over 120 thousand million million ways. So we can take it as read that Gamlin didn't do an exhaustive trial of all the possible choices.

Even after he had decided to place low numbers next to high ones, to punish inaccuracy, he still had plenty of choice. We can assess how severely any order penalises a poor throw by adding up the *differences* between successive pairs of numbers as you go clockwise around the board. The pair (20, 1) contributes 19, then (1, 18) gives 17, (18, 4) gives 14, and so on.

Dartboard No's	20	1	18	4	13	6	10	15
Differences	19	17	14	9	7	4	5	13

2	17	3	19	7	16	8	11	14	9	12	5
15	14	16	12	9	8	3	3	5	3	7	15

For *any* ordering of the numbers, the bigger the resulting sum, the bigger the penalty for just missing a high number will be (on average). For the standard board, the total of these differences is 198, but is it possible to get higher?

As it happens, the maximum difference total is 200, only two more than on Gamlin's board. In other words, the standard order of numbers on a dartboard *very nearly* exacts the maximum penalty available.

In fact, it is very easy to arrange the darts in a way that gets a difference total of 200. Call the numbers {11, 12, . . . , 20} the *High* numbers, and {1, 2, . . . , 10} the *Low* numbers. If you then alternate High and Low numbers around the board, the difference total will always be 200 – it doesn't matter how the High and Low numbers are ordered amongst themselves!

If there were any thought of disturbing Gamlin's ordering – which there isn't – then any High–Low board as we've just been describing would be worth considering. The dartboard below is an example.

You'll see that the bigger the number is, the smaller its neighbours are, so it does a decent job of penalising those who are too greedy and aim higher than their ability merits.

This board has a major drawback, however – it is too easy to make sure you get an odd (or even) number when you want to. At the end of a game of 501, there is a premium on being able to hit an odd number so as to leave an even number for the final double. To make the dartboard as challenging as possible, even and odd numbers should therefore alternate.

Is it possible to get the 'perfect' board that has alternating odds and evens, and also has alternating High and Low numbers to maximise the punishment for just missing?

The answer is no, and it is easy to prove why this is so. Suppose we start creating such a board at 20, which is an even, High number. Its two neighbours would therefore have to be odd, Low numbers. Their neighbours, in turn, would have to be even, High numbers, and it follows that for this dartboard to work, all the High numbers would have to be even, and all the Low numbers odd – which is impossible.

In other words, the final dartboard has to be a compromise, and Gamlin's seems to be as good as it gets . . . unless you know better.

Finishing the game

In the early stages of a single game of darts, the aim is simply to score as many points as possible, but in the closing stages, the darts player increasingly has to think about how to finish.

It is very rare to see top players pause to think during their throw. It is as though the relevant sums are hard-wired into their throwing arms. Not only do they step up to the 'ockey (or oche, as it is now officially spelt) knowing exactly what score they seek from each dart, they also have a fallback plan. From a score of 112, they aim to put the first two darts into treble 20, then single 20, and finish with double 16; but if the first dart should hit single 5, the instant fallback is treble 19, then bull.

Fallback strategies like this can be the making of a decent darts player. For example, suppose you are left needing 51 points from three darts. One way to achieve this is to aim at 17, followed (if successful) by double 17. This puts great faith in your ability to hit the double 17 notch, however. If, as is quite possible, you aim for double 17 but your dart lands just inside and you score single 17 instead, then you won't be able to go out with your third dart.

A professional would adopt a more robust approach. In this case he would aim for 19 first, reducing the target to 32 (double 16). If he then

Dartboard finish quiz

- *In between 170 (the highest possible three-dart finish) and 159 (the lowest target with no three-dart finish) there are four other scores where three-dart finishes are possible. What are they?*
- *What is the highest score at which a player has a choice of finish?*
- *What is the highest score for which a player has a choice of finish and where one of the choices is strategically much better than the other (because it gives him room for error)?*

Answers at the end of the chapter.

missed his target with the second throw and scored single 16 instead, he would still have a possible finish on double 8. In fact this approach of trying to reduce the score to a power of 2 (that is, 32, 16, 8, 4, 2) is standard practice for amateurs and professionals alike. In each case, hitting the single instead of the double always leaves you with a possible double finish.

In the case of 32, there is the added bonus that 16 is adjacent to 8 on the dartboard, and therefore if you miss double 16 and hit any part of the 8 segment, there is still an opportunity to finish with your next dart.

Since aiming at treble 16 was also a good bet for the early stages, the tip from all this is that the typical pub player should ignore the convention of aiming for treble 20s entirely, and focus their entire game around 16s. Though instead of the familiar booming 'One hundred and EIGHT-EEE', the best he could now hope for would be 'One hundred and forty-FOOOUUURRR!!!'

The benefit of throwing first

Just as snooker players dream of clearing the table with a 147 break, so darts players would love to score 501 using the minimum possible number of darts, nine. One way is to begin with seven treble 20s, and get the last 81 from treble 19, double 12. Score any single 20 at all, and the feat becomes impossible.

A nine-dart finish is an extremely rare achievement. Only a handful of players have ever achieved it live on TV. The odds of a top player achieving a nine-dart finish have been estimated at something like 1 in 1,500, so it's not likely to happen even in a whole championship. Which is perhaps just as well, because if all players played the 'perfect' nine-dart game every time, then the first player in any leg would always end up winning. He would reach 501 at the end of the third turn before his opponent had a chance to respond.

Even so, the player who throws first may have a considerable advantage when it comes to winning the leg, as we mentioned in Chapter 6. The best players typically score around 100 for three darts, taking four turns to reduce the target to around 100. They can expect to follow this with a couple of darts to set up the finishing double, and two more to complete the job. In other words, at the top level, sixteen darts is roughly the average number required to finish the game. At this standard, making a few reasonable assumptions about the probability of various outcomes (see the Appendix), the first player to throw can expect to win a leg about 65 per cent of the time. In a set with only three legs, the right to throw first in the first leg is very significant, and even in a five-leg set, the first to throw has a definite advantage.

Darts has recognised the unfairness of allowing a coin-toss to confer this significant advantage on one player. The normal way to eliminate luck is to let the players' skills decide who throws first – each player aims a single shot and whoever is nearest the bull has the first throw.

Note that this concern over the advantage for the first thrower only applies to the better players. For most mortals, a sixteen-dart finish is just a dream. By the time you are flinging your 37th dart at double 1, whatever nominal advantage you had from throwing first was lost some time ago.

Dartboard finishes – answers to the quiz

The scores between 159 and 170 from which a three-dart finish is possible are 160, 161, 164 and 167.

The first score at which a player has a choice of exits is 164 (60, 54, bull or 57,57, bull). The former is probably marginally preferable because each dart will be unimpeded, but there's no other strong reason for choosing one over the other.

The highest score at which a player needs to think strategically about a finish is 130. Here he has a choice, but only one option has a fallback strategy. If he goes for the option treble 16, bull, double 16, then just missing treble 16 scuppers his chance of finishing that turn. If, however, he goes for treble 20, 20, aiming to finish on a bull, then if he accidentally lands on single 20 with his first dart, he still has the fallback option of treble 20, then bull.

8

CHASING THE ACE

The best servers should practise their returns

The year 1967 was a big one for tennis. On 1 July, the BBC began colour transmissions, and decided that the honour of the first ever broadcast should go to Wimbledon. It was felt that a screen filled with green would have quite an impact on viewers, and so it proved.

Something else, though, arrived with less of a splash in 1967 but was to have an even greater impact on the future of the game. Wilson, the sports equipment manufacturer, introduced the T-2000. This steel-framed tennis racket, later made popular by Jimmy Connors among others, was the first serious challenge to the wood-framed racket, which had been in use for over one hundred years. It started a trend that had revolutionised the game by the mid 1980s. New materials meant that professionals could hit a ball far more powerfully than had been possible before.

Today, thanks to the technology behind the modern tennis racket, the game is dominated by powerful hitters, and nowhere is this more significant than in the serve. Numerous men now match the feats of the legendary Bill Tilden and Roscoe Tanner, who are believed to have exceeded 140 mph using old-style wooden rackets. With modern timing equipment, Andy Roddick in 2004 became the first player to be officially recorded at over 150 mph. Venus Williams has closed in on 130 mph, making her serve faster than many of the top men in the wooden racket era.

In top-class tennis, being the server is a considerable advantage, and as the power of the serve has increased, this advantage has grown. With surprising predictability, the top players win about two-thirds of the points against each other on their serve, at least on grass where the surface is fast. (For example, in the 1999 Wimbledon Ladies' final Steffi Graf won 67 per cent and Lindsay Davenport 69 per cent, while Goran Ivanisevic and Patrick Rafter won 68 and 69 per cent respectively in the men's final two years later.)

But how does winning the point convert into winning the game?

The advantage of a good serve

If you win a certain proportion of your service *points*, what proportion of service *games* should you win? Will the proportion of games that you win be the same as the proportion of points?

The simplest case is when both players win the same number of points on average, i.e. when the server has no advantage whatsoever, winning 50 per cent of the points in the long run. In this case, a service game becomes like a game of heads or tails, and each player will have an equal chance of winning the game. (If this were not so, it would be like saying that somebody calling heads has a better chance of winning than somebody calling tails.)

The chance of winning the game when the server has an advantage is rather harder to calculate, but if you make the simplistic but reasonable assumption that the server has exactly the same chance, p, of winning every point, then there is an exotic formula from which the chance of winning the game, G, can be calculated:

$$G = \frac{p^4 - 16p^4(1-p)^4}{p^4 - (1-p)^4}$$

If you want to know more about where that formula comes from, take a look in the Appendix, but otherwise, just admire it for all its powers of 4.

To work out the chance of winning the game, *G*, you have to plug values for *p* into the formula. If *p* is 1 (i.e. the server wins every point) then *G* is also 1, which makes sense, since winning every point is bound to lead to a win every game.

What about the situation for top players who, as we've noted, win about two-thirds of the points on their serve? Feeding $p = \frac{2}{3}$ into the formula gives the answer that *G* is just over 85 per cent. In other words, the server's advantage on each point has led to a significantly greater chance that she will win the game.

The results for each different value of *p* can be plotted in a graph, which is shown below. The graph shows how the chance of winning the game increases as *p* increases. The interesting feature is that it is not a straight line, but a curve. And the fact that it is a curve has a bearing on tennis coaching, as we'll explain in a moment.

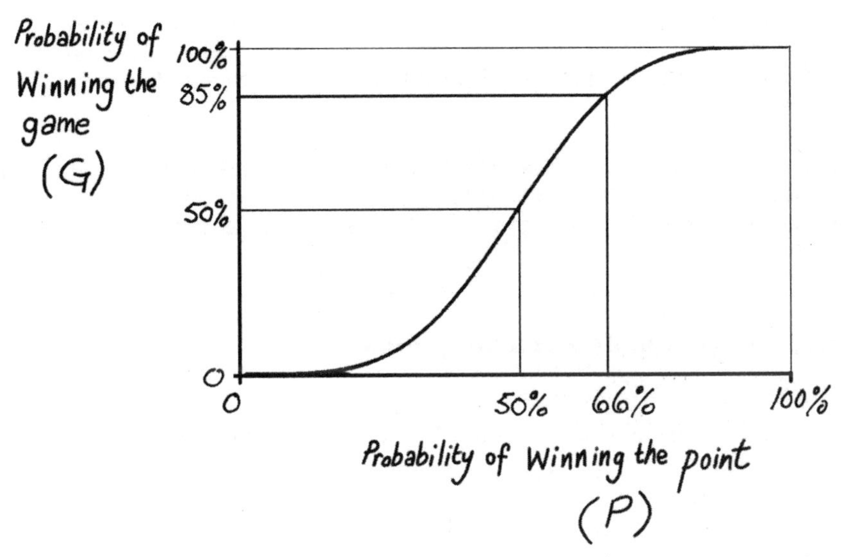

The slope of the curve near to where the player wins just half the points is steep – in fact for every 1 per cent by which the server increases his chance of winning a point, he increases his chance of winning the game by about 2.5 per cent. So, improving your serve percentage from 50 to 52 per cent would improve your chance of winning the game to 55 per cent. However, because the curve flattens the further you are away from $p = \frac{1}{2}$, the amount of benefit from a given service improvement falls away too. If a player could win just three-quarters of his service points, he would already expect to win 95 per cent of service games, and there would be little to gain from tuning his serve even more. Economists call this the Law of Diminishing Returns.

This leads to a practical tip if you are a respectable tennis player looking to improve your game. If you typically lose as many service games as you win, then time invested in improving your serve will pay handsome dividends. On the other hand, if your serve already wins you a large majority of your points, then devote your training time to other skills.

An ace puzzle

(To solve this puzzle, you need to be fully conversant with the scoring system for tennis.)

You are at home with the TV in the background and the sound turned down. You notice that Andy Roddick is in the middle of a match against Tim Henman and Roddick is serving. He serves an ace and goes on to follow this with five more aces, during which time Henman doesn't touch the ball once.

At this point, you turn the sound up, and hear that, despite losing all these points, Henman is currently ahead in the match. What is the match score at this point (in sets and games)? Answer at the end of the chapter.

Tennis scoring and crucial points

A server needs to gain at least four points to win a tennis game, though the scoring system does a good job of disguising this. Instead of calling the points 0, 1, 2, 3 and 4, tennis points have acquired the obscure labels 'love', 15, 30, 40 and game.

This scoring system has its origins in France, where the number 60 was always a dominant part of the counting system. (To this day, French counting comes to a curious halt at sixty. The French call the number seventy *soixante-dix*, or 'sixty-ten'.)

It is thought that the original tennis points were represented as the four quarters of sixty, and might even have been displayed using the hands on a clock. The first point was 15, followed by 30, 45 and then 'game' at midnight. Over the years, laziness led to 45 being shortened to 40. Meanwhile 'love' is usually thought to be a corruption of *l'oeuf*, the egg, because the number zero has the shape of an egg – though there are other theories.

The other crucial part of tennis scoring is that the game can only be won by a player who is two points ahead of his opponent. By the time the game gets to 40-all, one player now needs to win the next two points to win the game, a situation known as *deuce* (which is linked to the French word *deux*, or two).

Incidentally, if the scores in a tennis game reach 30-all, the umpire could fairly call deuce at this point, since the situation is exactly the same as at 40-all. But it would be very disconcerting to everyone if this happened, so the present system will stick.

There's a cliché among tennis players that every point is important, but it would also be true to say that some points are more critical than others. The crowd will almost certainly have an intuition for which points are the most vital. If a strong server leads 40–0 on serve then he won't be too worried if he loses the next point. On the other hand, if he loses a point with the score at 0–30 he is suddenly faced with three break points, and the crowd will gasp accordingly.

At what score within a game is the next point the *most* important? A complete answer really depends on the overall proportion of points a player wins. However, if you agree with the definition of 'importance' offered below, then Pancho Gonzalez, the best player of his era, was wrong when he argued that the most important point was when he was 15–30 down on serve.

A sensible way to assess the importance of a point is to measure how much difference winning or losing that point makes to winning or losing the game. And using this definition, 30–40 is always more important than 15–30. The reasoning is simple: if the server wins in either situation the result is, effectively, deuce; on the other hand, if the server loses at 30–40 he loses the game, whereas at 15–30 he still has a chance.

A second chance

There aren't many examples in sport where a player automatically gets a second chance, but that is what happens with a tennis server.

With very few exceptions, tennis players have two different types of serve in their repertoire, a fast serve (F), and a slower, spinning serve (S).

As they are allowed two serves, they have four possible tactics, which in shorthand we will call FF, FS, SF, SS. The standard tactic is to use F followed by S, though some players – Goran Ivanisevic was one – sometimes use FF; social players often choose SS. But there is an excellent mathematical reason why you never see the tactic SF, no matter how erratic the fast serve is, or how feeble the slow one.

Let's use some simple round numbers to illustrate. Suppose that in top-level men's tennis, the chance that a fast serve is good (i.e. finds its target) is around 50 per cent, and when that serve is good, the server then has an 80 per cent chance of winning the point. (These figures aren't too far from reality.) For the second serve, the server must be extra careful not to fault, and the figures indicate that a slower serve will be good at least 90 per cent of the time. When it is good, the winning proportion drops, let's say to 50 per cent. The figures in women's tennis are similar, though the fast serve is less decisive than the men's.

The chance of any given fast serve winning the point will be 50% x 80% = 40%. If it comes to it, the chance of a slower serve winning a point is 90% x 50% = 45%.

So if only one serve were permitted, the slower serve would be preferable because it wins slightly more of the time. But with two serves, look at each of the four tactics in turn. The chance of winning the point comes from two steps: first, find the chance that the first serve is good, and wins the point; then add to it the chance that the first serve is a fault, but then the point is won on the second serve. This leads to:

	First serve in and wins		First serve fault, second serve wins		Total chance of server winning point
For FF	40%	+	50% × 40%	=	60%
For FS	40%	+	50% × 45%	=	62.5%
For SF	45%	+	10% × 40%	=	49%
For SS	45%	+	10% × 45%	=	49.5%

FS has come out best here with 62.5 per cent, SF is worst with 49 per cent. And as it happens, FS will *always* be higher than SF, so long as a player's fast serve (when good) is more likely to win the point than their slow serve. If you want to see the algebra, look in the Appendix.

If 60 per cent of fast serves are good, rather than 50 per cent, the chance of winning the point with a fast serve exceeds the chance on a slow serve. So such a player should actually be advised to adopt the Ivanisevic FF tactic all the time. Only when the fast serve becomes less reliable *and* less potent (say 50 per cent for each) does the SS tactic come out top – as you will witness on any Saturday afternoon in the local park. So depending on the situation, FS, FF or SS can all be the best strategy, but never SF.

Of course, the very fact that nobody would expect a fast serve to follow a slow one means that a player might just try it as a surprise tactic. It might work once . . .

Some people claim that since having the right to serve is an advantage, why should players have two chances? There are many counter-arguments. At a social level, having the serve can be a handicap as much as a help, and the same rules should apply to these matches and those of Roger Federer. Moreover, in any match, the rules are the same for both players, so advantage or handicap will cancel out. If there were no second-serve rule, we spectators would seldom see players using powerful, but risky, fast serves – the game would be poorer for their demise. (The great Rod Laver suggested a compromise: allow second serves, but only for four points in each game. Second serves would become equivalent to playing the joker.)

And if second chances work in tennis, how about offering them in other sports? In soccer, if a team gets a free kick in a dangerous position, or wins a corner, why not add to the game's excitement by giving them a second chance if the first fizzles out? There would have to be some means of deciding when the 'first' attempt was deemed to be over. The benefits would include more goalmouth scrambles, more chance of reward for attacking play. On the other hand, there are probably enough arguments with the referee as it is.

Answer to the ace puzzle

The puzzle hinges on when it is possible to serve six consecutive aces in a tennis match. There is only one such scenario. The players were involved in a tie-break. Roddick won the last two points of the tie-break on his serve to win the set. At the start of the next set it was Roddick to serve, and he served four aces to win the game. We therefore know that Roddick won a set and was 1–0 up in the next set. But we are also told that Henman was ahead in the match. This can happen only if Henman has already won two sets. So the match score must be that Henman leads by two sets to one, Roddick leads 1–0 in the fourth set.

9

FIRST TO THE FINISH

The calculations of a sprinter

At the Athens Olympics of 2004, Kelly Holmes became the first British woman for 84 years to win two athletics gold medals. It was headline news, and for a couple of days the analysts in the studio could talk of little else.

Kelly won both the 800 and the 1500 metres using similar tactics. In both races she hovered at the back of the pack for the early stages, keeping out of trouble and running at her own pace, and then on the final lap she made her move to the front, striding past all the runners to snatch gold on the final straight. There was a risk to this strategy, however. To overtake in a middle distance race, you need to get past all the runners who are hugging the inside lane. This means you need to move out by a lane or two, and since doing this means you are running round a larger circle, you have to run further to cover the same distance (as the Red Queen might have said to Alice).

The fact that Kelly chose to run further than all the other competitors was clearly exercising the minds of some TV viewers. During one broadcast, the BBC presenter Steve Rider was handed an email, which read: *'I've noticed that Kelly Holmes has been running most of her races in the second lane. How much further does this mean she had to run?'*

Rider appeared to blush for a moment, then said, 'Well, I'm afraid I'm not the mathematician,' and turned to his co-presenter Sue Barker for help.

Like a relay runner, Barker immediately passed the question on to the American athlete Michael Johnson. Uncharacteristically, the great man failed to anchor the team safely home: 'I'm not sure about this, but I think in the second lane you run about 2.5 metres more,' he offered.

The correct answer is that running a complete lap one lane wide adds just short of eight metres, so Johnson's estimate was rather low. In the case of the 800 metres, if she ran the equivalent of both complete bends in the second lane, Kelly Holmes added almost 1 per cent to her distance, and over a second to her race time. Let's see where this calculation comes from.

Staggered lanes

It's a fact of athletics life that for distances longer than 100 metres and shorter than 26 miles, competitors have to run in circles. This has been the case ever since ancient Greece, when the first athletics stadia were built. One such circuit has been restored, the Panathaneic stadium in Athens, and it made a stunning backdrop to the Olympic Games of 2004 and also to the first modern Games in 1896.

The Panathaneic stadium is extremely long and thin by modern standards. In fact the straights are just over 200 metres long (about double the modern size) but the track is only 33 metres wide (less than half its modern counterpart). For ancient Greek athletes this must have posed a particular problem, especially for those drawn in the inside lane, because the radius of the curve is only about ten metres. The athletes must have been practically running on one leg as they leaned into the bend while running flat out.

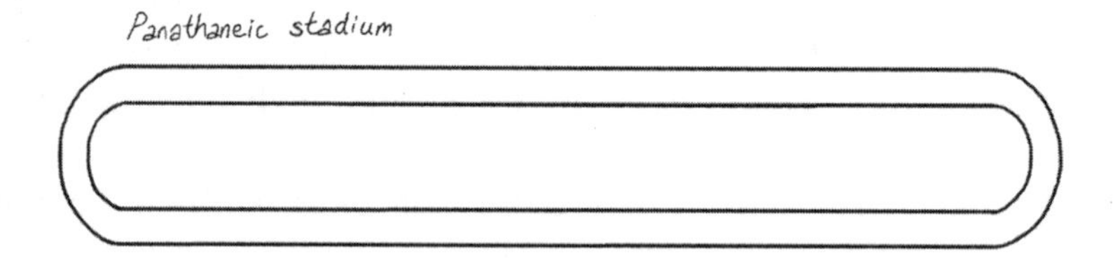

The fact that runners have to go around corners introduces the need for a staggered start, for the obvious reason that if all the competitors began the race level with each other, then those who had to run around the outer curve would have to run considerably further than those inside them.

You might think that the dimensions of an athletics track are specified to the last centimetre, but this is not the case. Because there are always limitations on the amount of land that is available, the authorities allow a certain amount of flexibility. So long as the distance around the complete inside lane circuit is 400 metres, the architects can make the straights relatively long and the curves tight, or the straights short and the curves more gentle. In practice, most stadia have straights that are about 85 metres long, and the curve's radius is about 35 metres. Each lane is between 1.22 and 1.25 metres wide (these ugly numbers presumably being the metric conversion of lanes that were once neatly four feet wide).

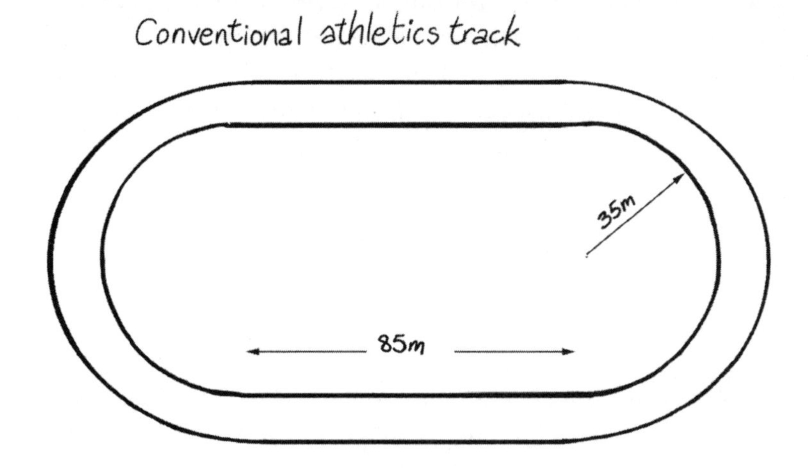

One constraint on the dimensions of the track is the javelin event, which takes place in the middle of the arena. The world record javelin throw is close to 100 metres, and the authorities allow for a 30-degree angle of error in the direction that the thrower might release. If the stadium is too circular, there is a risk of the javelin overshooting onto the track at the far end, and if it is too long and thin, the javelin might stray onto the track at the side. An athlete being skewered by a javelin might have been crowd-pleasing entertainment for the Romans, but it's not the done thing with twenty-first-century compensation lawyers on the prowl. It was OK to use the long thin Panathaneic stadium for archery (a more accurate sport) but not for the javelin.

The extra length of an outside lane can be calculated using the familiar equation for a circle. The circumference of a circle, you will recall, is pi (roughly 3.14, or 22/7) times its diameter. So to work out the distance around the curved part of an athletics track, multiply its diameter by pi.

The curved part of the second lane has a diameter just two lane widths, about 2.5 metres, more than the inside lane. So the *extra* distance running entirely in lane two is pi times 2.5, or about 7.85 metres. This is as true for the Panathaneic stadium as it is for Olympic Park in Sydney, and was the answer Michael Johnson was looking for.

Tactics

For distances up to 200 metres, runners strive to achieve top speed as quickly as possible, and keep going to the finish. But as many great athletes have demonstrated, running at full stretch for the whole race is not necessarily the best tactic for middle distance running. You will race as fast as you can at the very end, but what about initial speed, and where should your final sprint begin? In simple terms, there are four tactical mistakes that a middle distance runner can make:

- Start too fast, so fade at the end
- Start too slow, so too far back to catch up
- Begin the final sprint too early and lose speed at the end
- Delay the final sprint too long.

The 800 metres is a good distance to think about: it is long enough for tactics to play a role, but short enough to offer a limited number of options. The tactics are described by just two numbers: first, the time taken to cover the first lap; second, the distance from the finish at which to begin the final sprint. What combination will work best?

This question has been addressed by an American statistician, who persuaded a number of club runners to take part in an experiment. By using information from all of them, he was able to predict the total time for the race from the two numbers that were input. At the end of the experiment, it became clear that being able to organise yourself to run the first lap at a uniform pace to a definite target time had the biggest influence in reducing the total time for the run. Athletes might describe this as 'running your own race' – do not react too early if a rival carves out a big lead, and do not be afraid to stick to your pace if the others choose to lag behind.

We don't think Kelly Holmes' coach had read this particular analysis beforehand; but she ran her races exactly as this maths would advise.

The cat-and-mouse nature of 800 metres running is nothing when compared with the cycling race of the same distance. The 800 metre cycle sprint is one of the most bizarre sporting spectacles of all. For the first 600 metres, the two cyclists creep around the track, carefully eyeing each other like teenagers hanging around the streets on a Sunday afternoon. Then, with 200 metres left, they suddenly release all their power in a desperate bid for the finishing line. Only this final bit is really a sprint.

But why the long drawn-out trundle beforehand? The reason is that for most of the race, the cyclists both want to be last. This reverse logic is due to the effect of the slipstream. A cyclist meets plenty of wind resistance when hurtling at up to 50 miles per hour, but if he can tuck himself behind another cyclist, the wind resistance is considerably reduced. To conserve energy for the final sprint, it is therefore in both cyclists' interests to be in second place. The result can be a ludicrous standstill where both contestants demonstrate their abilities to balance on a static bicycle.

All-round winners

In several Olympic racing sports, there have been competitors who achieved success in more than one event. Kelly Holmes, Jesse Owens and Emil Zatopek all won golds in different athletics races, but these pale in comparison with the swimmers. Although Ian Thorpe and Michael Phelps have been the successful swimmers of recent times, perhaps the greatest feat of all was Mark Spitz's seven golds in 1972. (This produced an amusing example of commentary spin from the BBC. After several lengths of the 1500 metres freestyle came the words '. . . and the British boys are letting Mark Spitz make the pace *at this stage . . .*')

Nobody will dispute that to win one gold is a splendid achievement, and that to win two or more is remarkable. Yet there is something unsatisfactory about the fact that multiple golds are almost always won in disciplines where there are several similar contests, but over different distances. The correlation between similar events reduces our admiration for the subsequent successes.

After all, if a swimmer can win the 100 metres freestyle by a clear margin, it's not altogether surprising that he can win the 200 metres and even the 50 metres. Take the argument to an extreme: we could see a 100 metres race, a 120 metres race and so on, and award a gold medal in each. The same couple of swimmers would likely share all the golds between them, adding massively to their countries' tally.

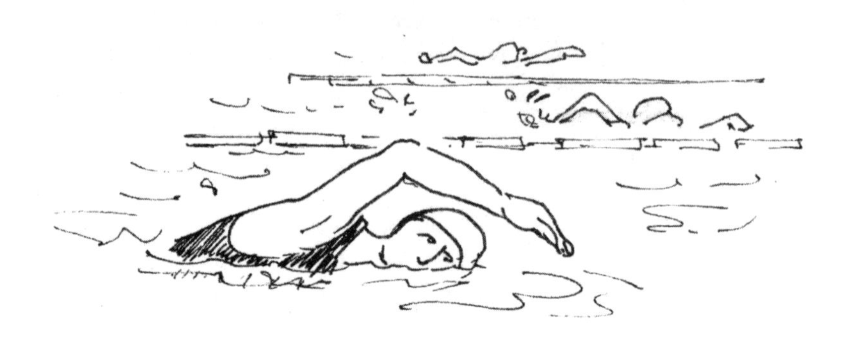

In contrast, multi-discipline events are almost worthy of more than one medal. It seems rather unfair that the men and women who win the decathlon and the heptathlon get only one gold medal each. These competitions are spread over two days, and are widely regarded as the best tests of who is the leading all-round athlete.

If they are to be fair tests, then the best sprinter should be no more and no less likely to win than the best jumper or the best thrower. The governing bodies strive for this ideal: for example, when fibreglass came into use, dramatically increasing the heights jumped in the pole vault, the scoring system was soon adjusted to bring the specialist vaulters back into line.

Nevertheless, proficiency in some events often goes with proficiency in others – top sprinters are often splendid long jumpers, and good discus throwers can usually put the shot well. To achieve *fairness*, it doesn't really matter if one of the events consistently tends to score even 100 points more than the rest, as all competitors get the same benefit. But it is important that the *variability* of the scores across the events is similar. If one event tends to display much more variability than the rest, then the specialists in that event will be favoured. The reason is that those who are best in that event will score more *extra* points (above the event's typical score) than those who are best in events that show less variation.

A study of five decathlon world championships reached clear conclusions. Except for the 1500 metres, the variability of performance within the field events was rather more than in the track events.

The 1500 metres is always the last event, and by the time it starts, most competitors will know that they are well out of the medals. This reduced incentive to press to the limit might partly explain why the performances in it are so variable. So if we set aside this last event, it looks as though the decathlon set-up favours the throwers rather than the runners.

This same study confirmed what you might expect. Athletes achieving high scores on the shorter track events tended to score highly on the long jump; and the scores for the three throwing events were strongly associated. But there were also some surprises: good throwers tended to be good pole vaulters, while the best high jumpers were often strong in the 1500 metres.

If you had to use performance in just one of the ten events to predict the winner of the decathlon, which would you plump for? You are looking for an event with 'transferable skills', and the answer might change if the scoring system changed. But at the moment, the answer is the long jump, just ahead of the discus throw. So if any decathlon coach is scouting for talent, he might hang around the long jump pit, and take the best jumpers for a try-out with the discus . . .

10

GOALS, GOALS, GOALS

The patterns that can predict results

Part of the romance of football, and the reason why it is so popular, is that while the great clubs like Real Madrid and Juventus tend to dominate, in any particular game there is always a chance of an upset. As the cliché goes, there are no easy games in football. A weekend with no 'shocks' would be a real surprise. It's just that we don't know in which matches they will come.

The main reason for the potential upsets is that football is a low-scoring sport. In rugby, a score of 30–19 would not be unusual; in basketball it might be 74–68; in one-day cricket, the teams may score 258 and 235. And in football? You can expect about 20 per cent of all the games you see to produce one goal at most. No wonder there is such a roar when the ball does cross the line.

Some of the clichés in the analysis of goals in football are suspect to say the least. If you ever hear the words: 'What a good time to score a goal!', ask yourself when might be a bad time to score. There isn't one.

And here's another. Are teams more vulnerable just after they themselves have scored? You will often hear this claim, though there is no real evidence for it. Of course you will recall those times when the opponents equalised straight after you scored, but a cold look at the figures shows that the fact that you have scored does not, in itself, make an opposition goal appreciably more likely.

When do goals occur?

Anyone seeking to understand or predict the development of a football game must ignore the folklore and examine the facts about when goals are scored, and by whom. Although the result of any particular match will be

> ***Manager maths***
>
> *Managers have been known to make occasional mathematical comments, some more insightful than others. Here are three from managers who shall remain nameless:*
>
> *I am a firm believer that if you score one goal the other team have to score two to win.*
>
> *Of the 100 youngsters I see on the training ground today, 99.5 per cent will be from this country.*
>
> *I want to turn this team around 360 degrees.*

gloriously unpredictable, we can use these figures to make pretty accurate predictions about how many homes, aways and draws will happen over a period of time, and what scores will be most frequent.

The chart shows the times of all the 10,409 goals scored in some 4,000 matches between 1993 and 1996 by the 92 clubs in the main English leagues. So you'll see that, for instance, nearly 400 goals came in the first five minutes, and about 600 between the 60th and 65th minutes.

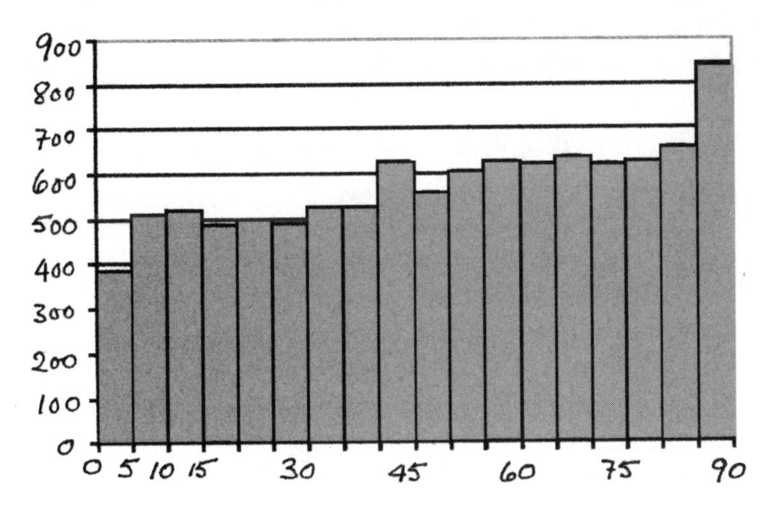

The 'blips' seen in the periods 40–45 minutes and 85–90 minutes are easy to explain. Any goals scored during the 'stoppage time' a referee adds on at the end of each half are *recorded* as having been scored at 45 or at 90 minutes. So these two time intervals are actually a bit longer than five minutes – there isn't really a rush of goals in the last five minutes of either half. It's also nice to see the figures confirming what we would expect – fewer goals in the first five minutes of each half than later on, as both teams will take a little time to get the ball and their attackers into scoring positions.

The chart shows a small but definite increase in the rate of goal-scoring over a match. About 4,600 goals were scored in the first half, while the second half had over 25 per cent more. And there is another small effect that statisticians have noticed from more detailed studies: the more goals that have already been scored in a game, the more we expect later. Goals beget goals – but for both sides: it is almost as though defensive slips and accurate shooting are contagious.

Home advantage

There is another important pattern in the study of goal-scoring, that of home advantage. Home teams score more goals and win more matches than away teams. This is true for all countries and all standards of football.

The two charts show the contrasting performances of home and away teams over the 30 seasons from 1972 to 2002 in the main English leagues. Home teams scored an average of about 1.5 goals per game, while for away teams the average was about 1.1 goals per game.

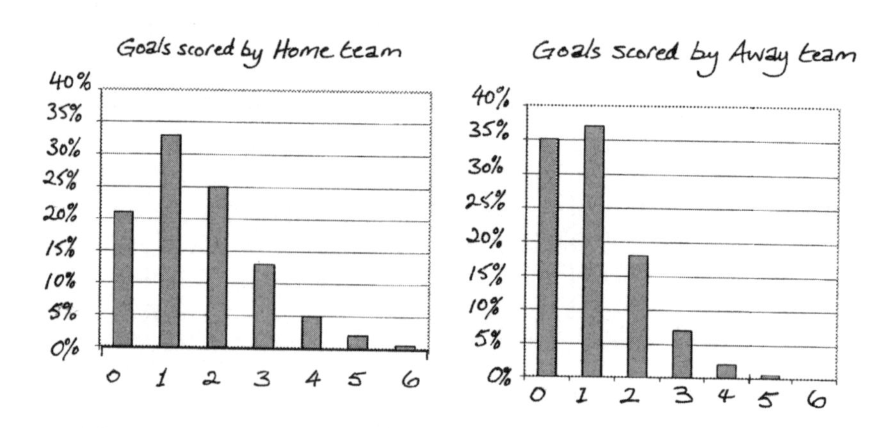

If goal-scoring is a random process, then these charts can be used to predict the chance of a particular result. What, for example, would be the chance of a 1–1 draw? From the away chart, you'll see that the chance of scoring one goal is about 37 per cent. Also, the home team scores one goal in around one-third of all its games. So if both teams score at these rates, we can estimate the chance of a 1–1 draw by simply multiplying these figures together. One-third of 37 per cent is roughly 12 per cent, which is around one match in eight.

In fact, using this same idea, we would predict that the following results are the most common in football, in order of likelihood:

1-1 Draw
1-0 Home win
2-1 Home win
2-0 Home win
0-0 Draw

And how does this prediction compare with real scores in top-flight football? It turns out to be spot-on. These are the five most common results, accounting for more than half of all games, and there's not an away win among them. And the real frequencies of these five results are close to the figures we get from using the charts.

The football goal model

To predict the most common football results, we used the separate figures for how often the two teams score exactly 0, 1, 2, etc., goals, as shown in the charts. But there is an even simpler way to estimate the chances of the various match outcomes, using only the average number of goals each team scores, and it turns out to be almost as accurate.

The observation that, by and large, goals are scored *at random* means that the actual number of goals a team scores should follow a statistical pattern, known as a Poisson distribution. Knowing just the average numbers of goals for home teams and away teams separately, this standard bit of statistics leads to their respective frequencies for exactly 0, 1, 2, etc., goals. And these figures are almost exactly the same as those shown in the charts. It really is enough just to know these averages.

To sum up, we offer the following broad description of a football match. Temporarily ignoring which teams actually scored, goals at football come along largely at random, at a rate of about one every 35 minutes; a tad more slowly in the first half, a little bit faster in the second, and also slightly faster when we've already had some goals. Within that picture, home teams average around 1.5 goals, away teams average close to 1.1.

And that broad description fits well with thirty years of match results. For any individual game, you can tweak the estimates of the average number of goals each team might score to take account of 'form', injuries, league position, change of manager, and so on, and hence assess the chances of a score like 3–1 *for that game*. But just using the overall averages for the numbers of home and away goals, the 'model' we offer for a match lets us make accurate predictions for the frequencies of individual scores, and then for match results.

Taking all possible scores together, the model suggests (and the actual results confirm) that about 46 per cent of games will end in a home win, 27 per cent as an away win, and 27 per cent as a draw. Home advantage is real and substantial, but the figures can't tell us why. Is it the roar of the home crowd, familiarity with the arena, or maybe subconscious bias from the referee?

Predicting the result

There is one group of people in the football world who could benefit directly from a result-predicting formula. When rain or snow causes a few matches to be postponed, the 'Pools Panel' is called up. Its job is to forecast the results of the postponed matches, for the sole purpose of allowing the betting on football pools to continue as though the matches had been played.

The Pools Panel are a group of ex-footballers and other experts who gather together in secret. Exactly how they come up with their decision that Notts County will beat Bury, or Aberdeen will draw with Hibernian, remains a secret, but no doubt voices are raised as the relative merits of the teams are debated.

Rather than shut a group of people away for a couple of hours, we could turn to computers. For any match, feed in the data of the *average* number of goals each team is expected to score, and simulate the matches in the manner we've described, goals for each team coming along at random for a period of 90 minutes. Then announce the 'result'. If the computer produces the score Gillingham 2 Crewe 1, then the Panel enters it as a home win.

Incidentally, the same idea can be used for any particular match you think of betting on, just from estimates of the average number of goals each team might score. If past form suggests Gillingham will score 1.8 goals and Crewe 0.9 goals, you can calculate the chance for each team winning and for a draw and then compare this with the odds being offered by the bookies. If there's a significant discrepancy in your favour, it may be worth a bet.

Of course, we are only dealing with averages here. When the fixture is played, Crewe might go on to demolish Gillingham by four goals to nil. But at least if it was a computer that had made the forecast, it would be the machine that took the blame for getting it 'wrong', rather than the panel who used it.

What happened next?

Meanwhile, how does all this help the fan? Our goal predictor can help to forecast not only what the result will be before the game begins, but also how a game that is under way is likely to end up.

For example, if your team scores the first goal, common sense says that you are more likely to win the match than lose it. But *how much* more likely? Common sense will say it depends on many things, not least at what time the goal was scored. If that first goal comes in the last minute of the match, then victory is a near certainty, whereas a very early goal gives plenty of time for the opposition to turn things round.

Suppose your team, City, is in a crunch derby match, with both teams level on points in the league. In derby matches, home advantage doesn't seem to count for as much, so we'll take both teams as equally likely to score at any time. You haven't heard the result yet, but thanks to some contrived circumstance, you earlier picked up the tantalising titbit of information that at one point in the match, your team was 1–0 up. Trouble is, you don't know when that goal was scored. How good should you feel?

Try our simple model. Suppressing the details, the maths suggests you should be pretty content. *If City went 1–0 up at one point in the match, then you should expect them to win that match about two-thirds of the time. There's only a one in seven chance that your team lost after scoring that first goal.* Even better, that's what real figures from real matches show as well.

As with so many questions of probability, you have to read the small print, because a tiny change in the wording will change the answer. For example, if the snippet that you had picked up from a match report was that 'City scored the first goal' it's safe to bet that there was more than one goal in the match. (Otherwise the reporter wouldn't have used the word 'first'.) And the more goals that are scored, the less likely it is that scoring the first will prove decisive. Hearing the snippet 'City scored first' is therefore worse from your point of view than hearing that City were one up at some point.

And here's a different snippet that you might pick up from the report on the TV news: you tune in just in time to see City scoring a goal, and then they move on to another story. You know City scored, and you know it was the *last* goal in the match (they always show all the goals in sequence). But what is the chance that City *won*?

According to our model of a match (which is looking pretty reliable), the answer should be exactly the same as for when they score the first goal. Just *run time backwards*: the last goal becomes the first, and even if the rate of scoring goals changes a little, the matter of which team scored any goal remains the same. You ought to win two-thirds of the games when you score the last goal, the same proportion as when you score the first goal. And that's what happens in real life.

This idea can be taken further. What if you score the *second* goal – are your chances of winning the match still about two-thirds? No! You are *less likely* to win matches in which you score the second goal than matches in which you score the first! The maths suggests that you'll win around three-fifths of matches in which you score the second goal.

The reason for a reduction is subtle but simple: the very fact that you scored the second goal means that at least two goals were scored during the match. Any games with just one goal are not included in the figures. But obviously, in all these eliminated games, the team scoring first got the *only* goal, and so won the match. Even *asking* who scored the second goal disposes of a number of games, in all of which the team scoring first were victorious.

The important influence of the first and second goal on the result is all down to the overall low level of scoring in football. In rugby or basketball the first score is, in the grand scheme of things, much less significant.

The red card trade-off

Not only can the pundits and the fans benefit from the football predictor model. It might be of help to players too, albeit for a cynical reason.

Football managers have been heard to moan that it is 'harder to play against ten men than against eleven'. Any such claim is quite a comment on their own tactical inadequacy, because common sense tells you that it must be false. Otherwise, teams would constantly be removing one of their own men. But how big an advantage is it to have the extra man? How many goals is the extra player worth? One reason why a football player ought to know this is that he may have to make an instant decision – should he, in the interests of his team, commit an offence that will get him a red card if, by doing so, he stops a probable goal?

This is a genuine dilemma only in close matches, so let's assume the scores are level. We surely also need to know how long there is left to play: in the last minute, the (cynical) player will always accept being sent off to stop a certain goal, else the match is lost, but whether he should do the same with 20 minutes left isn't obvious.

Let's ignore those times when committing the red card offence would also lead to a penalty kick: that will nearly always be a bad idea. To come up with the advice below, three Dutch statisticians looked at what happened in 340 games where just one team had a man sent off. This enabled them to estimate the average extra number of goals that would be scored, according to how much time remained. Now they could do a balancing act: on the one hand, the chance of a goal if the nasty foul is not committed, on the other the likelihood of shipping extra goals if you must play with a man short.

The key thing to identify is the *crossover time* in the game. Before that time, your team are better off if you act the gentleman and let play continue; but after that time, your team are better off if you commit the foul and get sent off. The crossover time will vary depending on how likely a goal is to be scored if you don't foul.

Deep breath: if you consider the opposition are absolutely certain to score, then do the dirty deed as early as the 16th minute. But we've all seen top strikers miss sitters, so taking such an early bath will happen very rarely. If the scoring chance is about 60 per cent, then don't commit the foul before about 48 minutes. And if the chance is as low as 30 per cent, hold fire until the last twenty minutes of the game.

11

ELEVEN (AND OTHER ODD NUMBERS)

Numbers and numerology in sport

If you play a team sport, then the chances are that you will find yourself playing against the odds. We're not talking about betting here, but about the curious fact that in most sports, teams are made up of an odd number of players. In fact, we have scoured our sports encyclopaedias, and it's hard to find any major worldwide sport where teams are not odd. Start with the best-known international team sports:

Football 11
Cricket 11
(Field) Hockey 11
Rugby League 13
Rugby Union 15

Add to these the three major American sports:

Basketball 5
Baseball 9
American Football 11

And in addition, teams in many of the smaller sports are also odd – hurling and Gaelic football (15, down from 21 or more in the early days), speedball and bandy (11), netball, water polo, kabaddi and handball (7), and so on.

There are exceptions. There has to be an even number of oars in a boat or it would tend to go round in circles (though the presence of a cox ensures that the number becomes odd again). Polo has four in a team, volleyball and ice hockey are both played with six. But if you play in a team sport, the odds are high that your team will be odd.

Elevenses

What is noticeable in the list of team sports is the popularity of the number eleven. Is there some significance in this number? It could be argued that eleven is quite pretty. It is symmetrical (upside down or in a mirror, it's still eleven) and it also has some curious mathematical properties. $11 \times 11 = 121$, $11^3 = 1{,}331$, $11^4 = 14{,}641$ – all of them palindromes. If you take any number of the form abcabc, for example 851,851, then it will always be divisible by eleven. And if you are looking for really obscure elevenisms, then how about this: Eleven + Two is an anagram of Twelve + One. Now that must mean something!

Properties like this will certainly have appealed to mystics and numerologists in medieval times, and could have helped to give the number eleven some symbolic meaning. But would this be enough to persuade the administrators of the newly formed association of football to choose eleven as a team number? Probably not. After all, there are also downsides to the number eleven. It is a prime number, which means there is no way of dividing it up into equal groups; unlike twelve, say, which can be divided into sixes, fours, threes or twos.

Odd occurrences of the number eleven

If you look hard enough for any given number in sport, you can find it all over the place. Here are some more instances of the number eleven:

- *There are exactly eleven different ways that a batsman can be out at cricket;*
- *In recent years, the scoring system of table tennis has been changed so that the target to win a set is now eleven points, and not 21 as hitherto;*
- *It takes eleven officials (one chair umpire and ten line umpires) to oversee tennis matches on the main courts at Wimbledon. We know of no other sporting contest where the ratio of match officials to players (11 to 2 in tennis singles) is so high.*

The first team sport to choose eleven as its official number seems to have been cricket, when revised Laws for the game were published in 1835. Until that time, cricket teams had often been made up of eleven or twelve players, apparently at random. And if you go back far enough there was almost no regulation at all. The first Laws of cricket said nothing about the number of players in a team, and fixtures where, say, eighteen Gentlemen played against Eleven of All England were not uncommon. After all, the essence of sport is a contest between near-equals, so handicapping the stronger team to have fewer players is fully in that spirit. But why *Eleven* of All England?

Perhaps a clue for the prevalence of the number eleven in cricket comes from other numbers that had been established in the sport in the 1700s. The official length of the pitch was decreed to be 22 yards, while the official height of the stumps would be 22 inches, both multiples of eleven.* Twenty-two yards is a very old measurement – it is one-tenth of a furlong, and the typical width of a Saxon farmer's strip of land, later to become known as a chain. Perhaps the Marylebone Cricket Club simply liked the fact that the number 22 was already a cricket number, and so decreed that the number of players in the match should also be 22.

Association football was next to select eleven as its official number. Many of the earliest football clubs were offshoots from cricket clubs (Sheffield United, for example), so the idea of eleven players in a football team was almost certainly borrowed from cricket. The fact that cricket-playing schools and universities such as Eton, Harrow and Cambridge were all involved in writing down the early rules of football would have reinforced this number.

Soon afterwards, the number eleven crossed the Atlantic. A team from Yale University took on Eton in 1873, and had their first taste of eleven-a-side football. When the rules of American football were formulated a few years later, it was Yale who influenced the choice of the number of players in the team. While most American colleges preferred rugby's idea of fifteen in a team, Yale pushed for soccer's eleven, because they felt it encouraged more open play. So, bizarre as it may sound, the reason why American football has eleven players is, at root, almost certainly because of cricket. How many Americans would guess that?

---

* These days stumps are 28 inches high.

There's another, more banal reason why eleven might have become the most common team number in sport. Eleven is one more than ten. A cricket team can be described as a captain and his ten men, or ten fielders and a wicket-keeper. A football team consists of ten outfield players plus a goalkeeper. In fact in almost all 'goal' sports, there is an odd player at the back, either a goalkeeper or a very defensive full-back, which allows the rest of the players to be divided up equally between attackers and defenders, and players on the left and right. This would help to explain the prevalence of odd numbers in team sports.

Or it could all be a coincidence. The reading of 'meaning' into numbers, or numerology, applies just as much in sport as in astrology.

All these numbers are 'man-made'. But sport itself can generate patterns. These will occur randomly, but it can still look as if there is some greater hand at work.

Lucky and unlucky numbers

When team sports introduced numbers onto the back of players' shirts in the 1920s, those numbers began to take on a personality all of their own. For some people, the number 9 will forever be synonymous with Bobby Charlton, and the number 15 with J P R Williams. But the significance of

Eleven World Cup winners

The eleven winners of the FIFA World Cup from 1962 onwards form an uncanny, almost perfect pattern of countries:

1962 Brazil
1966 England
1970 Brazil
1974 W Germany
1978 Argentina
1982 Italy
1986 Argentina
1990 W Germany
1994 Brazil
1998 France
2002 Brazil

Only England/France spoils the symmetry, and even they are neighbours in both their geography and the alphabet. Those who believe that diagrams like this have some deep significance use them to make forecasts. Brazil won the World Cup in 1958, so the pattern predicts their victory in 2006.

The edition of the book you are now reading was first published in 2005. If you are reading it after Brazil's World Cup victory in 2006, you will be struck by our prescience, and might be tempted to bet on Germany for 2010 (after all, West Germany won in 1954). If, however, Brazil failed to win in 2006, it will give you further evidence for the limitations of numerology.

shirt numbers is raised to another level in American sports, where certain numbers are retired in honour of the player who wore them. You'll never see a New York Yankee wearing the number 3, for example, because that was Babe Ruth's number, and it was retired from use when he died in 1948. And in the year 1999, the shirt number 99 (another multiple of 11) was retired from NHL ice hockey in honour of its wearer, Wayne Gretzky.

What about the number 13, traditionally regarded as the unluckiest number? You won't find the number 13 being retired from baseball teams very often, for the simple reason that tridecaphobia (the fear of the number 13) is so rife in American sports that players hardly ever wear it in the first place.

Outside the USA, this superstition is less pervasive, though it is not uncommon. Rugby league seemed happy enough to choose thirteen as the number of players in their teams, but Huddersfield, the place where league broke away from union, traditionally numbered their players as 1 to 12, and 14. The rugby union teams Bath and Richmond had a similar policy, and hence a number 16 shirt, while other teams bypassed the problem by using letters, until the authorities stamped out this traditional nonsense by standardising the shirt-numbering schemes, and squad numbers became the norm.

The bad luck associated with 13 can even attach itself to other numbers. Australian cricketers have traditionally regarded 87 as an unlucky score. Why? Because it is 13 runs short of a hundred. A batsman is therefore happy when his score moves past this number. And a score of 111 – known colloquially as 'Nelson' – also seems to be feared by cricketers.*

* It has been claimed that the name 'Nelson' comes from the (mistaken) belief that Lord Nelson had one eye, one arm and one leg. The number 111 also resembles the shape of a cricket wicket.

Needless to say, if a wicket ever falls when either number is on the scoreboard, the commentators will draw attention to the fact, whereas when the score passes without incident the non-event will quickly be forgotten.

But is there any evidence that a batsman is more likely to be dismissed on some particular score, rather than at other scores? To assess a batsman's vulnerability, or *hazard rate* at a particular score, say, just ten runs, two figures are required:

N = The number of times he ever scored *at least* ten runs, and
R = The number of times he was out for that score.

His hazard rate is then taken as the ratio R/N and the bigger this is, the more vulnerable he has been at that score. If there is any significance in the unluckiness of 13, 87 or 111 in cricket, then we would expect to see a blip in the hazard rate at those points.

Of course, such a calculation can only begin to be reliable if there is plenty of data. If a batsman never makes it to a score of 87, how can we know if he is vulnerable or not? One batsman who scored at least 111 runs enough times to begin to form a view was Don Bradman, the greatest batsman who ever lived. But even Bradman didn't play enough big innings to be able to distinguish reliably between his vulnerability on particular scores. For the record, he was never out for 87 or 111 in a Test match, whereas he was twice dismissed for 112. In fact, over the history of Test cricket to the end of 2004, only 43 batsmen were out for 111, compared to 61 who lost their wicket at 112. Perhaps the relief of getting past the final unlucky score leads to a lapse in concentration and the inevitable clatter of stumps.

Getting even

Although odd numbers seem to predominate in the size of teams and many of the other significant statistics, there are areas in sport where the even numbers fight back. The most obvious is that, in order to try to be fair, many contests are arranged so that both teams have the same opportunity to enjoy any advantages thrown up by nature. So in soccer, rugby and the like, the game has two halves, and the team that played with the wind/down the slope/at the muddy end in the first half plays the other way in the second half.

American football, and Australian Rules, go even further, dividing the game into four quarters, with teams changing direction at the end of each period. A full game of polo is split into eight equal 'chukkers'; a rule unique to this sport is that ends are changed every time a goal is scored! For indoor sports, the argument about nullifying the capricious advantages gained from nature has less force, but even here, almost all sports are divided into an even number of periods. The major exception is ice hockey, with its hour of playing time split into *three* 20-minute periods, making it 'A game of three halves' as one author put it.

Sometimes the notion of symmetry inevitably throws up an even number: golf courses that have an 'outward' and 'inward' series of holes will have an even number, almost always 18, in total. A croquet lawn, symmetrically arranged round the finishing pin, will have an even number (six) of hoops.

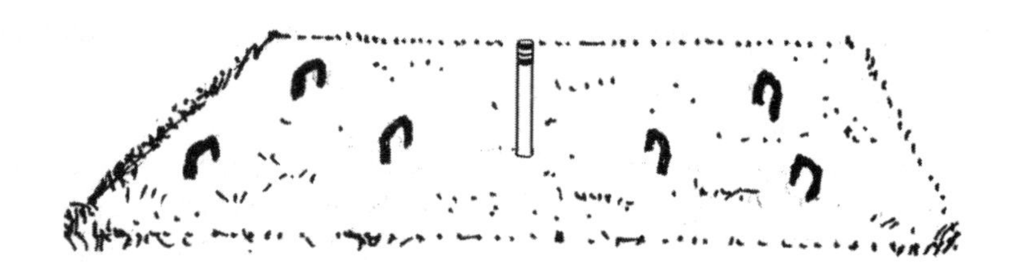

There seems no overwhelming reason for it to be so, but major athletics tracks and Olympic swimming pools have eight lanes, and relay races, on land or in water, tend to have an even number of legs, usually four. (However, the multi-event competitions, the bi-, tri-, pent-, hept- and dec-'athlons', do split among odd and even numbers.)

In almost all competitions that are organised into leagues, each league will have an even number of teams; and if teams meet more than once, they generally do so an even number of times, allowing equal amounts of 'home advantage'.

The organisers of knockout competitions love to work with powers of 2 – major tennis tournaments seed 16 or 32 players, and when the 'big' teams start to play in the third round of the FA Cup, just 64 contestants remain. If the total number of teams who enter such a competition is not a power of 2, then some preliminary round or rounds will be used to reduce the numbers appropriately.

Incidentally, a favourite 'trick' question in sports quizzes is to ask how many knockout matches need be arranged altogether, if (say) 373 teams enter. Don't try to work this out by asking how many preliminary matches are needed to get down to a power of 2, then how many more are needed. Realise that to produce the one winner of the competition, all the other teams must be eliminated: each match will eliminate one team, so the number of matches needed is always *one fewer* than the number of teams. Just say '372 matches, next question please'.

12

GETTING THE ANGLES RIGHT

Snooker and friends – an elliptical excursion

When it comes to precision, few sports rival snooker. Anybody who has crouched over a cue and aimed at the black at the other end of the table will appreciate that it can be quite an achievement even to hit the target, let alone send the black in the right direction. And unlike rifle shooting or other target sports, snooker throws up a huge diversity of situations, so the player has to learn where to hit the ball in many different positions, and with different amounts of strength and spin. On top of that, the player knows that the smallest of errors in any shot can hand the match to the opponent. At the very highest level, a player can lose a match 5–0 without playing at all badly.

It is possible to determine just how 'pinpoint' this accuracy has to be, by working out the amount by which the target ball is misdirected if you make a tiny error in the direction that you hit the cue ball.

From now on, the cue ball will be called the 'white' and the target ball the 'black', whatever colour is actually aimed at. (In a black-and-white book, it makes things much simpler.)

Suppose you hit the white at a slight angle to the black, like this:

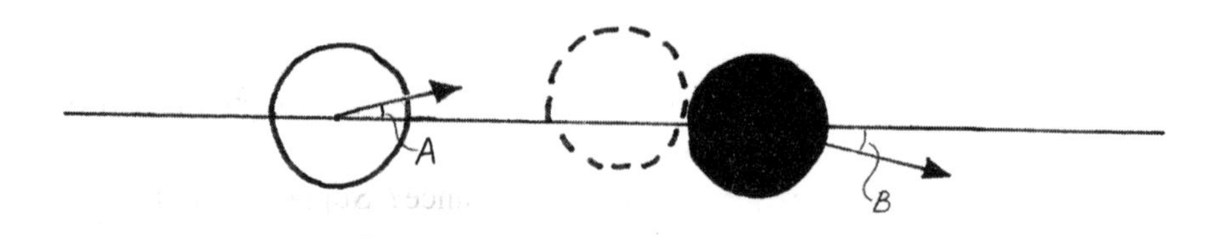

The dotted circle shows where the white strikes the black. Now whatever the angle at which the white is hit (we've called it A here), the angle B by which the black deviates will be bigger. This will *always* be true, as long as the gap between the white and black is bigger than 5.25 centimetres, the diameter of a full-size snooker ball.

And for shaky snooker players, this geometrical fact is the beginning of the bad news. Because what it means is that a slight error in the intended angle A will be magnified into a bigger error in angle B.

It gets worse. Suppose you miscalculate where the white should strike the black, and aim just a tiny bit away from the correct spot on the surface of the black. Let's call the error E, just for the hell of it. It turns out that if you double E, then you more than double the amount by which the black aims off target. In other words, the worse your error, the more magnified it becomes.

The diagram below shows what's happening:

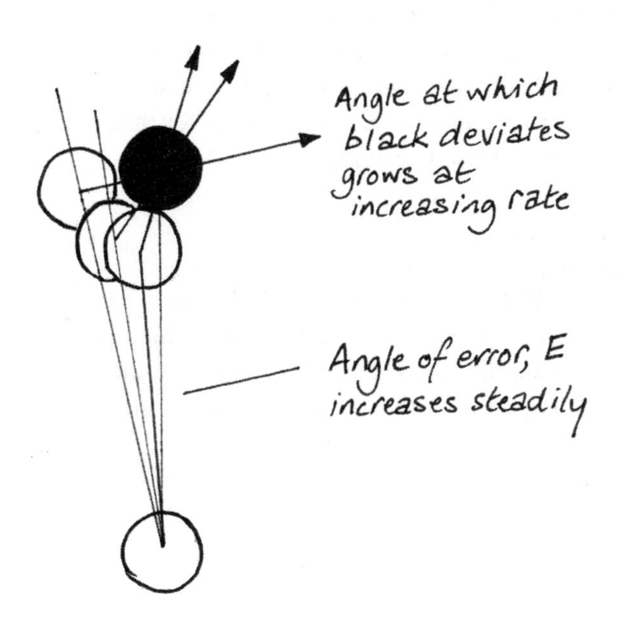

This begins to explain why snooker – and pool, for that matter – can be such fiendishly unforgiving games.

How big are these errors in terms of distance? Suppose the distance between the white and the black is about 50 cm (probably on the low side for many snooker shots). How does an error in the angle at which the player directs the white translate into an error in the black ball? The maths (or Sod's Law, if you want to view it that way) says that when the object ball is 50 cm away:

- If it is a dead straight shot – that is, if A is zero – then even at this short range, any error in straightness is magnified about ten times.
- And if you want to deflect the ball at 45 degrees, for example when attempting to pot a ball on the green spot into the near corner, with the white placed on the yellow spot, the magnification factor for an error is close to 15.
- And with the black ball on its own spot, a metre or so from the corner pocket, an error of even *one-third of a degree* in the white's direction will mean the black ball ends up three centimetres off target. That is quite enough to guarantee that it rattles out of the jaws of the pocket.

And the larger the original distance between the two balls, the more the error in the direction you send the ball is magnified. Double the distance between white and black and you more than double the error on the black.

All of this only increases our admiration for the consistent skill shown by the likes of Ronnie O'Sullivan.

The hardest shots

The one shot that doesn't require the player to work out any angles is the straight pot, that is, when the white and the black are perfectly aligned with the pocket. Suppose the white ball is a given distance from the pocket: ideally, would you prefer the black to be near the white, halfway to the pocket, or adjacent to the pocket?

The maths shows that the easiest of them all is when the white and the black are almost touching and aligned with the pocket. Even if you only strike the white vaguely in the right direction, the black will head almost straight for the pocket.

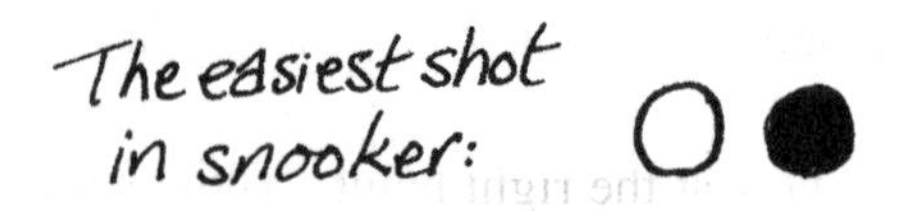

Almost as easy is the straight pot when the black is over the pocket. Here, as long as the white is hit straight enough to at least kiss the black, it should go in.

Still easy:

However, when the distances from the white to the black and the black to the pocket are both large, the straight pot becomes much harder.

Harder:

Once again, the problem arises because any error in the direction of the white leads to a magnified error in the direction of he black. It turns out that this error is at a maximum when the black is midway between the white and the pocket.

Which is the hardest pot of them all? Pressure, of course, has something to do with it – when Dennis Taylor crouched to pot the final black in the epic final of 1985 against Steve Davis, a routine shot must have taken on Everest-like proportions. But disregarding the psychological factor, the hardest shot will be the one where the player can afford the least margin of error.

Ignoring obvious constraints, such as either ball being very close to the cushion, there are two factors at play. The first is the error magnification factor that we discussed earlier. This suggests that the hardest shot will be a long distance, very fine cut: for example attempting to pot the black ball from its spot with the white somewhere in the D at the other end of the table. (No surprise, then, that snooker players never seem to take on this shot.)

There is another factor though: the player needs to know *where* he should be aiming the white to hit the black at the right point. (There's no point being accurate if you are aiming in the wrong direction!) Arguably, with a very fine cut you know exactly where to aim the white – you want it to graze the very edge of the black. But if you want to hit the black less finely than this, where should you aim the white then? Only a supreme head for trigonometry or years of experience will tell you.

What might be the hardest (feasible) unobstructed pot of all to accomplish, even without the complication of bouncing the black off the cushion on its way? We look for a combination of the four ideas that the white should be a long way from the black, the black a long way from the target pocket, the desired angle of deflection large, and the pocket as narrow as possible.

That suggests two general candidates. First, the two balls at opposite ends of the table, the black equidistant from either pocket – rather like the position when a black is respotted if the frame is tied. Secondly, the black about three-quarters of the way down one side of the table, a few inches away from the side cushion, needing the most precise of fine cuts to squeeze it into a centre pocket. The white will be near the opposite corner pocket. If you can succeed with either shot, you have a bright future in the game.

Where the white goes

Snooker is not just about potting. A player who wants to compile a decent break is just as concerned about where the white goes as he is about the ball he is trying to pot. This point is also a vital component of the game of billiards, where a player may be seeking to score points by glancing the cue ball off the object ball, either to make a cannon, or even to send the cue ball into a pocket. And, indeed, this consideration is at the heart of success in the ancient game of shove ha'penny, where you try to nudge an existing coin into a bed, and also hope the shooting coin falls into a bed.

The laws of mechanics give a good indication of which direction the white is going to head. If the balls are not spinning, and if no energy is lost when white and black collide (these are two big ifs!) then the white will head in a direction that is always perpendicular to the direction that the black goes. How far it goes depends on how hard it is hit.

The real world is not quite so simple, even with no spin imparted. If the white is *rolling* forwards, and not just skidding, that will give it a tendency to continue moving in the same direction. The overall outcome would be that the white is deflected at a smaller angle.

So two colliding snooker balls theoretically do this:

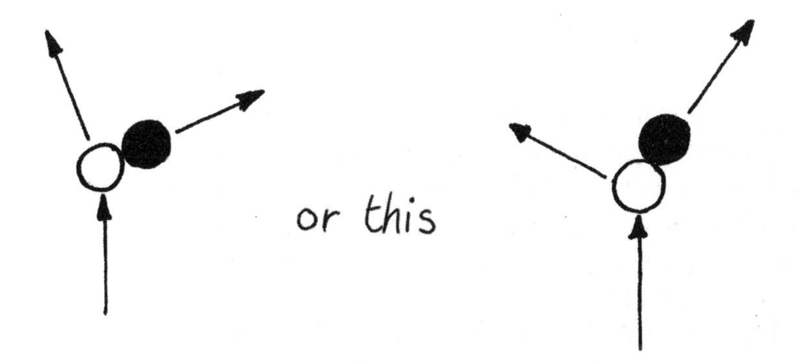

But in reality, with forward rolling, unless the player applies some spin to the ball, they are more likely to do this . . .

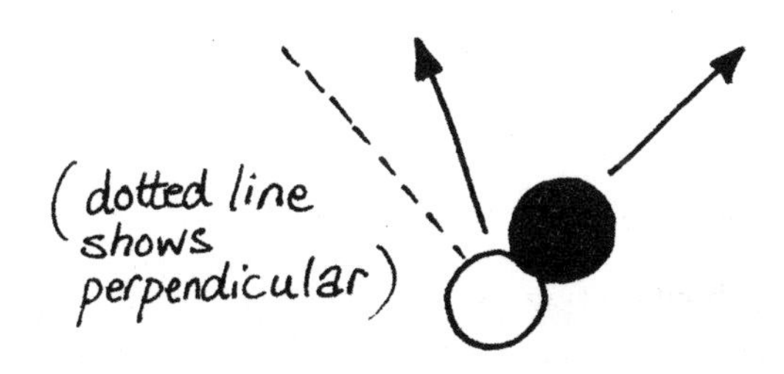

Defence can be an equally important part of the game, and one of the best moves is to place your opponent in a position from where they cannot make a direct hit on the object ball (i.e. a snooker). Escaping from a snooker can present the player with even more problems than attempting a tricky pot. It means either swerving the white ball, or bouncing it off at least one cushion, and with any luck ending up with the balls in a position from where the opponent can't make an easy pot.

The way that balls bounce off the cushions is similar, but not identical, to the way that light reflects off a mirror. In the case of light, the angle at which a beam strikes the mirror is the same as the angle at which it reflects. The beam of light from a laser placed inside a rectangle of mirrors would look something like this:

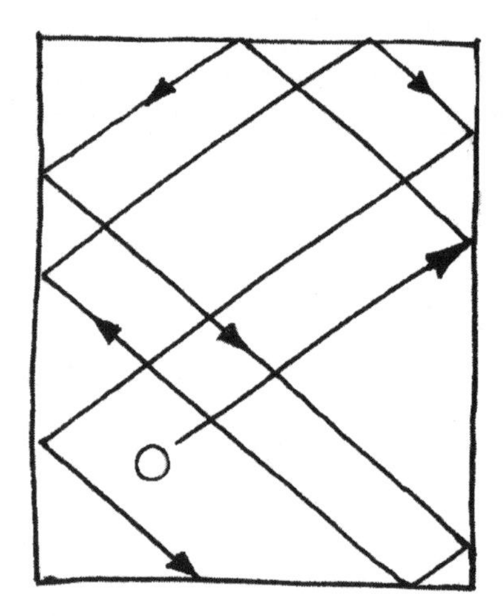

Notice how the beam always travels in parallel lines (either north-east or north-west), forming a pleasant pattern as it goes. This, in an idealised world, is just how the path of a snooker ball would look as it bounced off multiple cushions. It suggests that even if you are limited to only one direction where you can aim the white, you can still reach most regions of the table, if you bounce off enough cushions – and of course, provided there are no rogue balls or pockets in the way.

Real snooker tables are not a bad approximation of the mirrored room, but real balls encounter something that doesn't apply to light – friction. Imagine the white ball is struck dead centre, with no 'side' or spin imparted. When it hits the cushion at an angle, the friction or grip of the

A mathematician's 'trick' snooker shot

In the idealised world where snooker cushions behave like mirrors it is possible to get the ball to return to its starting point after hitting precisely four cushions. But there is only one way of doing it. Wherever the ball is on the table, you must aim it at a specific angle that is very close to 26.5 degrees from the long side of the table. The ball's path will then map out the sides of a parallelogram.

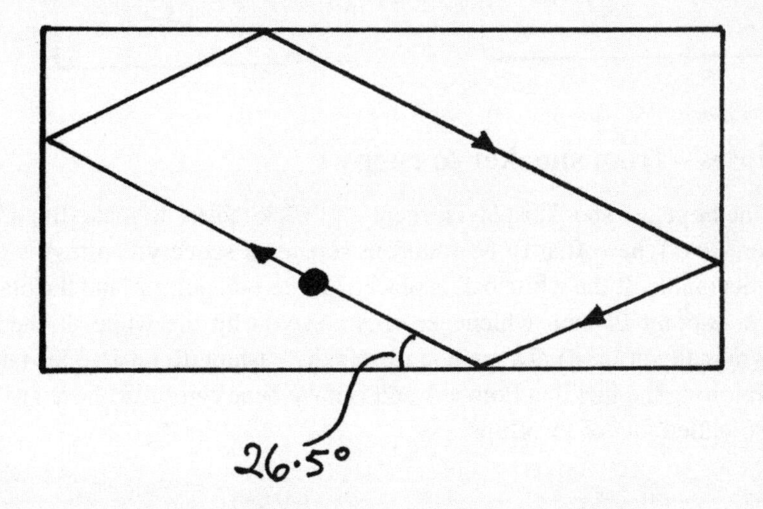

This angle applies to any snooker or pool table of regulation dimensions, i.e. the length has to be exactly twice the width. (The angle of 26.5 degrees is in fact the angle between the diagonal of a snooker table and its long side. If the table dimensions change, so does the 'magic' angle.) One way of testing out how close to the ideal a real table is, is to attempt this shot next time you find yourself at such a table. Remember to take a protractor with you. (You may get some strange looks from the bar, but that's life.)

cushion against the ball causes it to spin slightly. As a result, it will bounce off at an angle that is smaller than the angle at which it hit the cushion. The more acute the angle at which the ball strikes the cushion, the more noticeable this deviation becomes. It gives the appearance of the ball 'hugging' the cushion. To counteract this, the snooker player has to apply some reverse spin when he strikes the ball, if he wants the ball to bounce back at the 'correct' angle.

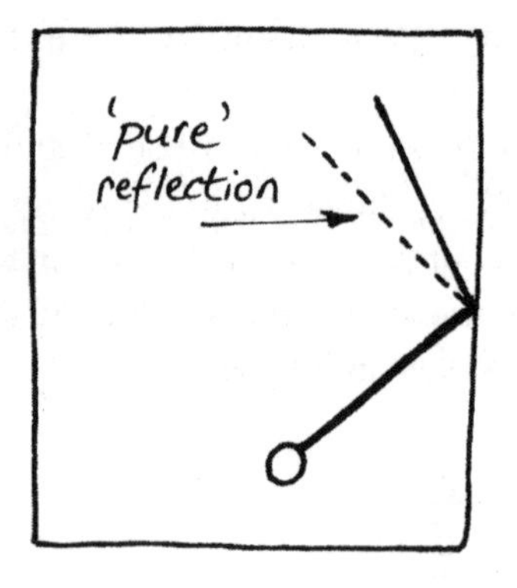

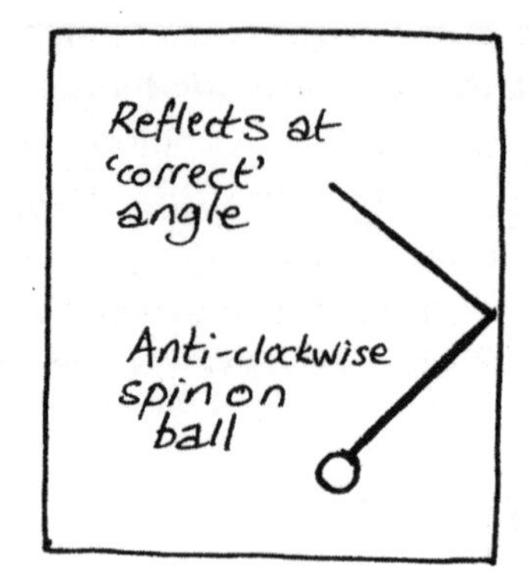

Ellipses – from snooker to rugby

For the hopeless snooker player, there are 'trick tables' to make life a bit easier. One type (often to be found in science discovery centres) is the elliptical table. If the white ball is placed at a certain point A and the black ball at a point B, then whichever direction you hit the white, it should (theoretically at least) always bounce off the cushion to hit B. This table is exploiting the fact that Points A and B have been carefully chosen to be the so-called *foci* of the ellipse.

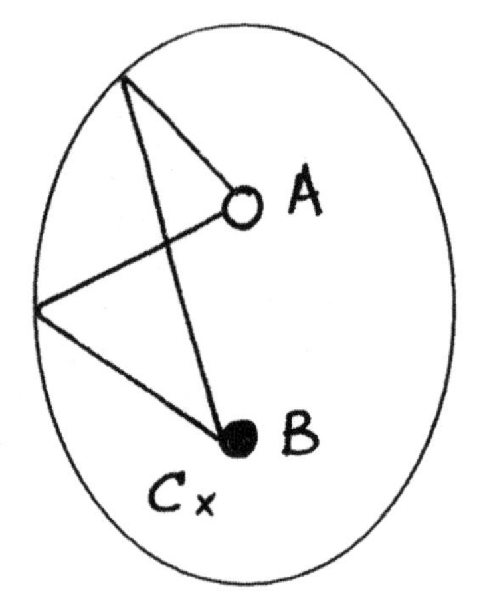

All roads lead to B . . .

While this is a wonderful way of ensuring that even a truly incompetent player can strike a white ball at A so as to hit the black at B, even if the direct path is blocked, it has its downside. Move the target from B to any other point, such as C, and you will be in big trouble. All your shots that hit the cushion will home in on B, and very few of them will be in the right direction to go on to hit C.

It's easy to draw an ellipse, using two drawing pins, a pencil and a piece of non-stretchy string. Attach the ends of the string to the drawing pins, and place the drawing pins on the paper at the points where you want the foci. Use the pencil to pull the string taut, and then move it around the drawing pins. The curve that the pencil marks is an ellipse.

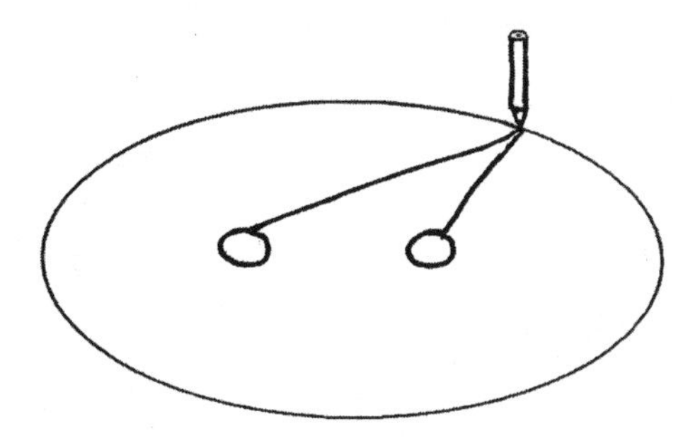

This construction shows one other property of elliptical snooker tables: when you do send the ball from one focus to the other via a cushion, the total distance the ball travels is the same, no matter what direction your initial shot.

And now that you've mastered the art of ellipses, let's make a seamless transition to another sport where ellipses also have a curious (if obscure) part to play. The sport is rugby, or more specifically, the kicking of conversions or penalties. As with snooker, the rugby kicker has to send a ball in a particular direction with little margin for error. After a try is scored, the kicker is allowed to attempt a conversion by placing the ball anywhere on an imaginary line drawn perpendicular to the goal line from the point where the ball was touched down.

If the try is scored between the posts, the only decision for the kicker is how far back to go, to ensure enough lift for the ball to clear the crossbar. But for a try wide of the posts, there is another factor – to make the angle subtended by the posts as large as possible.

If you have read the book *Why Do Buses Come in Threes?* you will recall that there is an optimal point to place the ball. Too close to the goal line, the kicker can't see any gap between the posts; too far away, and the two posts appear as narrow lines in the far distance. Here is another way of finding that optimal position.

Think of the two rugby posts as the foci of an ellipse. Mark the widest point of the ellipse at the point on the goal line where the ball was touched down, and then draw the rest of the ellipse.

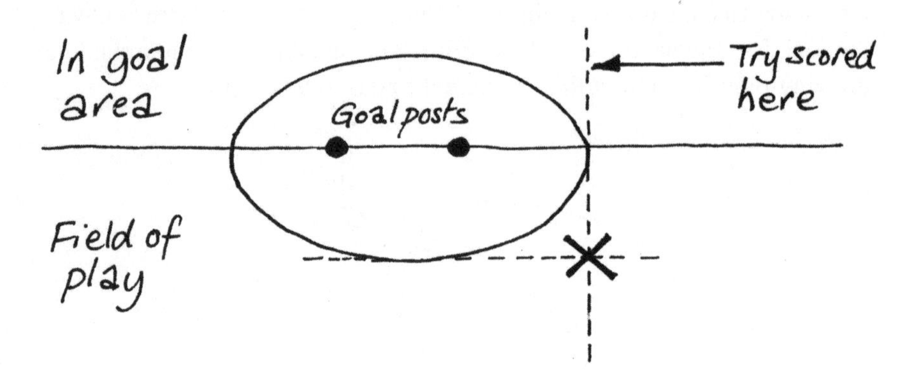

Next draw the tangents to the bottom and the side of the ellipse. The point where these two lines meet, marked X, is the point at which the kicker will find the biggest angle through which to slot the conversion.

There's a further curiosity that connects important mathematical curves with rugby. Mathematicians will know that if you slice a cone, the shape of the curve where the cone is cut will be a circle, an ellipse, a hyperbola or a parabola, depending on the angle of the cut. If you plot the optimal point for taking a conversion for every position wide of the posts where a try is scored, the resulting curve is a hyperbola. And when the kicker sends the ball on its way, the flight of the ball follows the path of a parabola. The idea of sending an ellipsoidal ball on its parabolic path from a point on a hyperbola is immensely satisfying.

13

TOP OF THE LEAGUE

The facts and foibles of league tables

We live in a world of league tables. There are league tables for schools, hospitals, and even celebrities. But the league tables that most people follow belong in sport.

The idea of a 'league' of teams competing against each other dates back to the 1870s, when the National Association of Baseball was formed. When it came to awarding the pennant, however, there was a snag. The winners were the team that won the most games, but since there was no regulation on how many times each team should play the others, the teams that played the most games were invariably at the top of the league. This meant that the pennant was as much a measure of keenness as it was of ability, though no doubt teams were keener if they won regularly. In 1872, the Boston Red Stockings played 47 matches, while the Washington Nationals only played 11. Unfair, until you realise that Washington lost all 11 games, and probably felt enough was enough.

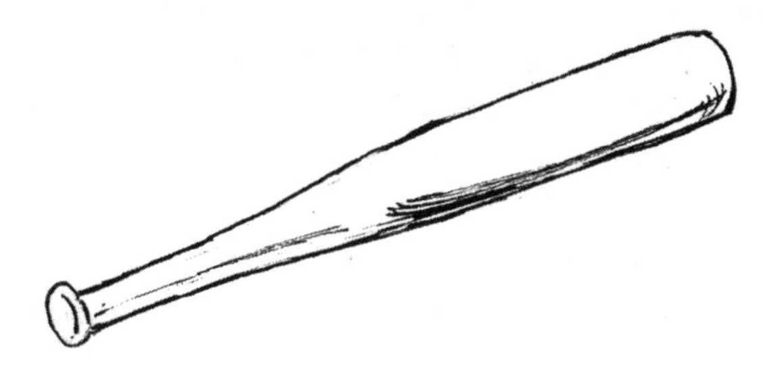

By the 1880s, some sports began to appreciate that each team in the league should play each other *the same number of times* (something we take for granted today), and also became aware of a need to find a sensible way of ranking the teams.

In baseball, the ranking was simple: the team with most wins was awarded the pennant. Ties were rare, caused only by bad weather conditions – the idea of a tied game has always been alien to American sporting culture. The situation was different in soccer, where drawn matches have always been common. Since a draw is halfway between a win and a defeat, the Football League agreed on the principle of two points for a win and one for a draw, and in 1888 they adopted a system that survived for nearly a hundred years.

Cricket's contortions

In the early days of cricket's County Championship, the rule was slightly different. For example, the first official championship in 1890 ended like this:

		PLAYED	WON	LOST	DRAWN	POINTS
1	SURREY	14	9	3	2	6
2	LANCASHIRE	14	7	3	4	4
3	KENT	14	6	3	5	3
4	YORKSHIRE	14	6	3	5	3
5	NOTTINGHAMSHIRE	14	5	5	4	0
6	GLOUCESTERSHIRE	14	5	6	3	-1
7	MIDDLESEX	12	3	8	1	-5
8	SUSSEX	12	1	11	0	-10

At first glance it seems bizarre that Sussex could end the season on minus 10. Did they have points deducted for some sort of felony? In fact the explanation is simply that points were calculated by taking the difference between games won and games lost.

It may not look like it, but this cricket table is directly equivalent to the football table adopted by the League two years earlier. For each county, if you increase its number of points by the number of games it played, you get the same outcome as you would in a football league awarding 2 points for a win and 1 for a draw.* The ranking is the same, it's just the psychology that is different. Nobody likes to end up in the red.

* In the cricket table, a win counted as one point, a draw as zero and a defeat as minus one. Add one point for each match and you get 2 for a win, 1 for a draw and 0 for a loss, the same as the football points.

Another look at the first cricket points table will reveal that the bottom two teams, Middlesex and Sussex, only played twelve games. Indeed, in the next season, the number of games played by the different counties ranged from twelve to sixteen.

If teams are playing different numbers of games from each other, the win-minus-loss system has its merits. It is certainly fairer than simply rewarding wins regardless of the number of games played. What it doesn't do, however, is discriminate between a team that wins none and loses five, and another team that wins five and loses ten.

Cricket took over fifty years to finally decide that all the teams should play the same number of games. Until then, it periodically fiddled with the points system to try to come up with something fairer than crude win-minus-loss. For a while it ranked teams according to their points divided by total number of games 'finished' (i.e. games that weren't drawn).

So, for example, in the 1896 season, Gloucestershire and Warwickshire both ended with minus 5 points, but Gloucestershire ranked higher because they finished more games (fifteen compared to Warwickshire's eleven).

COUNTY	P	W	L	D	Pts	Finished	Percent
GLOUCESTERSHIRE	18	5	10	3	-5	15	-33·3
WARWICKSHIRE	18	3	8	7	-5	11	-45·5

This is quite reasonable. Both teams were doing poorly, and if the unfinished matches had split pretty much the same way as those that did finish, Gloucestershire could expect a relatively better return from three more games than Warwickshire would from seven. But the authorities running cricket still weren't happy. For a start, some teams still had negative points, which didn't look good.

Cricket continued to change and tweak the system up until 1938, including the introduction of bonus points. Then Hitler invaded Poland. By the time Europe was at peace again, cricket had had a few years to reflect on the issue. All the problems over finding a sensible league table had come from the fact that teams played different numbers of games. After the Second World War, all the first-class county cricket teams played the same number of matches. Percentages were a thing of the past.

But the English summer was not long enough for each of seventeen counties to play all the others twice, and fit in other desirable games, so teams played only *some* of the other teams twice. This introduced bias, depending on whether you were lucky enough to play the weaker teams more often than the stronger ones.

Of course, all along everyone knew that the simplest and fairest system was the one used in football, with everyone playing everyone else once, at home and away. Finally, in the year 2000, cricket split into two divisions, and within each division every team now played every other twice. It took cricket a mere 110 years to find this solution.

Breaking ties

Whatever the method used for creating a league table, there is always the risk that two teams will end up tied on points. If they are competing for promotion or trying to avoid relegation, some method is needed to decide which team is superior.

There is no unique best method of breaking ties, but we can explore what rules have been used, and what their consequences are. In football today, the most common way is to use goal difference, calculated by

Two league table teasers

Football tables have been the basis of many a brainteaser over the years. These puzzles ask you to work out what the scores were in all the matches that have been played so far this season. They include relatively straightforward ones, like this:

Each team played the others once, what were the scores in each match (2 points for a win, 1 for a draw)?

	PLAYED	GOALS FOR	GOALS AGAINST	POINTS
UNITED	2	1	0	3
CITY	2	2	1	2
ALBION	2	0	2	1

. . . to stinkers like this:

The league table below got smudged in the rain, and is only partly legible. Eventually each team will play the others once, but the tournament isn't over yet. Can you find all the results of the games played?

	PLAYED	WON	LOST	DRAWN	FOR	AGAINST	POINTS
ATHLETIC	3	2	*	*	4	4	*
ROVERS	*	1	*	0	3	0	*
TOWN	*	*	*	0	1	1	*
WANDERERS	*	*	*	*	*	*	*

Both puzzles require a degree of deductive reasoning. One way to begin to tackle them is to realise that, for example, the total of the 'Win' column is always the same as the total of the 'Lose' column (one win for one team must create one loss for another team), while the total of the 'Draw' column will be twice the number of matches that have been drawn (because every draw involves two teams). Similarly, the 'Goals For' and 'Goals Against' columns must add up to the same total.

The solutions to the two puzzles are at the end of the chapter.

subtracting goals against from goals for. But until the 1976/77 season, the normal method for separating football teams level on points in a league was to use goal *average*, or goal *ratio* (the ratio of goals for to goals against), not goal *difference*.

This distinction might seem cosmetic, but its effects are quite different at the opposite ends of a league table. At the top, where we expect teams to have scored more goals over the season than they have conceded, the ratio of goals scored to goals conceded will usually be large: a record of 60 scored and 40 conceded has a goal ratio of 60/40 = 1.5, a goal difference of 60 – 40 = 20. A team with the same points total, who scored 80 and conceded 55, has an *inferior* goal average, but a *superior* goal difference. At this end of the table, goal difference tends to favour teams who score a lot of goals, even if they concede a lot too.

But at the bottom of the table, the position of a team scoring 40 and conceding 60 is *better* (under goal difference) than a team scoring 55 and conceding 80; but *worse* under goal average (check this!). For relegation, goal difference favours teams who don't concede many goals, even if they don't score many – the exact opposite of what happens at the top of a league table.

Mathematically, the use of goal difference has the advantage that an undefined calculation, such as 10/0 or 0/0, cannot arise. But the use of goal average has an advantage of a more subtle nature: it is more likely to actually separate the two teams! This is because goal difference is inevitably a whole number, whereas goal ratio takes fractional values – and in a typical league table, far more fractions are possible than whole-number differences.

There are other methods of discriminating between teams or players, variants of which tend to be used in mini-leagues, for example those in the first stages of World Cups. One of the cleverest is based on the Sonneborn-Berger (S-B) method, originally used for breaking ties in chess tournaments. Its application is quite simple: start with 2 points for each win, 1 for a draw. If your (initial) score ties with that of one or more competitors, your S-B score is found by adding up the initial scores of all the players you have beaten, and half the scores of the players you drew with. So this method gives higher weight to doing well against the better opposition.

Let's see it in action, when the results among four players are as shown. No match was drawn, which makes life a bit simpler. Each row shows how the player on the left performed against the other three opponents.

	Anna	Brian	Connie	Dave	Points
Anna		Win	Win	Lose	4
Brian	Lose		Win	Win	4
Connie	Lose	Lose		Win	2
Dave	Win	Lose	Lose		2

So in the initial table, Anna and Brian won two games (and are tied) while Connie and Dave both won one game (so are also tied).

The respective S-B scores are:

Anna beat Brian (4 points) and Connie (2 points)	= 6 points
Brian beat Connie (2 points) and Dave (2 points)	= 4 points
Connie beat Dave (2 points)	= 2 points
Dave beat Anna (4 points)	= 4 points

This successfully breaks both ties, as Anna gets ranked above Brian (whom she beat), but despite Connie beating Dave, Dave has moved ahead of her by virtue of his win over the top player, Anna.

Since the ties have been resolved, you could stop there – and most tournaments probably would. However, what happens if teams are still tied after doing these calculations? You can actually keep on applying the S-B system until all the points are different (this is discussed in the Appendix).

If this all appears to be getting complicated, then that's because finding a *fair* system that works in all cases is much harder than it might seem. Too often, the authorities setting up a league system develop rules they think are fair but which can backfire seriously. Some of these are described in Chapter 14.

Promotion and relegation

League tables have an extra edge when there is the prospect of teams being promoted or relegated between them. Until the 1960s, English football had four divisions, each with around 22 teams, with two teams promoted or relegated each season. County cricket had a single league, with no relegation. For many teams and their fans, this meant that the end of the season carried little excitement. Even with many games left, the majority of the teams in the league might have nothing to play for except pride.

Football moved first. In 1973/74 the number of teams that could be relegated from the top division was increased to three. Later, to add spice to the lower divisions, in addition to teams at the top gaining automatic promotion, four further teams could take part in a play-off. A team in mid-table could be just a couple of wins from the fringes of promotion, or a couple of losses away from a relegation fight, virtually all season.

When cricket opted for two divisions of nine teams, the promotion/relegation zone was even more dramatic – three up, three down. While all this creates excitement, there is a disadvantage. If too many teams can change division, it becomes something of a lottery as to whether the divisions truly reflect the best and worst teams.

As a rule of thumb, we suggest that the number of teams who qualify for promotion or relegation should be roughly one fewer than the square root of the number of teams in the league. (There is no great mathematical theorem that decrees this should be so, but it seems to work!) This would give the following numbers:

NUMBER IN LEAGUE	SQUARE ROOT	NUMBER TO RELEGATE/PROMOTE
4	2	1
9	3	2
16	4	3
25	5	4

If the numbers are higher than this, the leagues will tend to be too volatile, and decent teams will be relegated more often than they deserve. If the numbers are lower, the leagues will tend to become stale and more dull.

Of course, what matters most in a league is who comes top and who comes bottom – particularly if promotion and relegation are at stake. One question the pessimistic manager asks himself at the start of the season is *how many points am I going to need in order to beat the drop?*

The English Premier League currently has 20 teams, and each season one team will be champions, while the bottom three will be relegated. Is it possible to work out how many points are needed to avoid relegation? And how certain can we be that those teams who finish near the top or

bottom of the league are really performing differently from the rest – or could it all be a matter of chance?

In the nine seasons between 1995 and 2004, the points total of the champions ranged from 91 (Manchester United, 1999/2000) to 75 (United again, 1996/97). At the foot of the table, the unwanted record was held by Sunderland (2002/03, 19 points only), while West Ham might be thought the team most unlucky to be relegated – their 42 points in 2002/03 was insufficient, in contrast to Bradford in 1999/2000 who survived with only 36 points.

Up to 2004 at least, every team with 35 or fewer points was relegated, any team obtaining 81 or more points became champion. The season with the narrowest range of points (34 at the bottom, 75 at the top) was 1996/97.

We noted in Chapter 10 that around 46 per cent of all matches are home wins, 27 per cent are away wins, and 27 per cent are drawn. So suppose every single match in the season were just a random event, all teams equally good, with those three outcomes occurring with those respective frequencies. How many points in such a league would secure the championship? What points total would lead to relegation?

As a start, we can work out the likely 'mid point' in a league table. In a random match, the home team obtains 3, 1 or 0 points with respective chances 0.46, 0.27 and 0.27, and so will get:

$$(3 \times 0{\cdot}46) + (1 \times 0{\cdot}27) + (0 \times 0{\cdot}27) = 1{\cdot}65$$

on average. A similar calculation shows that the average number of points from its away games is 1.08, leading to a season's total of about 50 points. That sets a benchmark against which we can judge whether a team is performing better or worse than 'average' – we would expect the 'middle' team in the Premier League to get about 50 points. And that seems to tally with what actually happens in practice.

But what about the spread of points at the top and bottom of the table? Direct calculation is extremely difficult, but fortunately it is possible to obtain an accurate answer using computer simulation. It is relatively straightforward to program a computer to use its built-in random number generator to simulate a complete season's matches using the above parameters, and produce a league table. An extra couple of programming lines gets the job repeated 1,000 times, and gives us an excellent idea of the *variability* to be expected from season to season. This method of working out the likely spread of outcomes by running a simulation many times is known as *Monte Carlo Analysis,* so named because of the random way roulette numbers arise in the famous casino.

Over our 1,000 computer simulations, the champions obtained, on average, 67 points; 90 per cent of the time, even the champions did not get more than 72 points. Be clear what that means: if all the teams in the league are equally good, every match being decided by random chance, some (random) team will come out on top, and it will score about 67 points – maybe fewer, maybe a little more. So if, in the real world, a team did win the Premier League with a total of around 67–70 points, it would NOT have convincingly demonstrated that it was better than any other team at all – including perhaps some of those relegated! But the data shows that the top teams in the League are obtaining totals rather higher than this. The champions do perform significantly better than would happen under random chance.

At the other end of the computer-simulated leagues, the team finishing bottom scored 37 points on average, while the averages for the other two relegation positions were 41 and 43. It was quite rare for the third-bottom team not to get at least 40 points. That gives a clear warning: failing to achieve 40 points is prima-facie evidence that a team is deserving of relegation, even though it might be lucky enough to find three worse teams. As the real data shows, the teams near the bottom are obtaining so few points that they cannot convincingly attribute their position simply to bad luck.

But any manager who is sacked because his team were relegated, *despite* scoring 41 points, can feel hard done by. Three teams were always going to be singled out for the dreaded drop. There is an arguable case that it really was bad luck, and not lack of competence, that sent this team down. Try telling that to the directors as they contemplate the future of the relegated manager.

Answers to the two table teasers

First puzzle:
United beat City 1–0
United drew with Albion 0–0
City beat Albion 2–0

Second puzzle:
Athletic beat Town 1–0
Athletic beat Wanderers 3–1
Rovers beat Athletic 3–0
Town beat Wanderers 1–0

14

ADDING TO THE THRILL

Ways to make sport more exciting

You might think that the aim of a televised sporting contest is for the best side to win. But the truth is rather more complex. Yes, we viewers expect the contest to be fair and for justice to be done, but we want much more than that. We want the contest to be *exciting*. When Michael Schumacher won seven races in succession on his way to his fifth consecutive Formula One Grand Prix championship in 2004, he got plenty of adulation, but the authorities were distinctly worried. The sport had become boring.

As the demands of TV audiences have grown, the TV companies have increasingly looked to make the sports they are covering more dramatic. It is a TV producer's worst nightmare to have a dull contest pushing beyond the allotted time schedules, forcing all the other programmes to be rescheduled.

Two factors in particular help to boost the ratings:

- The guarantee of plenty of 'critical' moments within a match; and
- Contests between the top teams or players.

And by clever (or sometimes not so clever) tweaks to the rules, the organisers strive to increase the frequency of at least one of these two factors.

Take tennis, for example. If a statistician were asked to design a match to assess which of two players was more proficient at tennis, he would count all the points equally, and whoever won most would be declared the winner. This would be the most 'statistically efficient' approach, but it would also be dreadfully dull.

Suppose the target was first to 100 points, and Venus Williams was leading 80–60. At this point, Williams would be content to rely on her powerful serve to coast to victory. The match would effectively be dead.

Whether they realised it at the time or not, the inventors of the tennis scoring system found a way of making the contest far more exciting than this. They achieved it by breaking up the match into games and sets. Each game has at least one critical 'game point' and each set has at least one 'set point', which is even more critical. In a sense, it doesn't matter how well the player does in all the other points, she has to win the critical points if she is to register on the scoreboard. It is theoretically possible in a tennis match to win over 40 per cent of the points and never win a game; 55 per cent of the points and never win a set; and an incredible 65 per cent of the points in a five-set match and yet still lose the overall match. (Can you figure out how? One way is described at the end of the chapter.)

Victory in tennis really is about the ability to win the 'big points'. And there are lots of them, too.

Table tennis

Table tennis is one of the sports that has learned from tennis. If you haven't played it for a few years, you probably remember taking it in turns to serve for five points, with the first to score 21 points winning the set, and the winner being the first to three sets.

All that changed in 2001. Overnight, the sets were shortened to become the first to 11 points (with a lead of at least two points) and matches became the best of (typically) five or seven sets. The reason for this change? To increase the level of drama in a sport that in most countries struggles for prime-time TV coverage.

Under the previous rules, a typical score in a tight table tennis set was (say) 21–18. So, if the better player found himself 6–2 down he wouldn't fret unduly, knowing there was plenty of opportunity to recover. But under the new rules, 6–2 down is pretty serious – to win the game without having to get through a tie-break at 10–10, he must win at least nine of the next twelve points. Thus the early points in a set have added importance.

But this fact is partly offset by another: if the sets are half as long as before, there will be twice as many of them, so the importance of each set is diminished. Being 1–0 down in the best of three sets is more desperate than being 2–0 down in the best of seven. So although the proportion of important points *within a set* has increased, any given set has less importance.

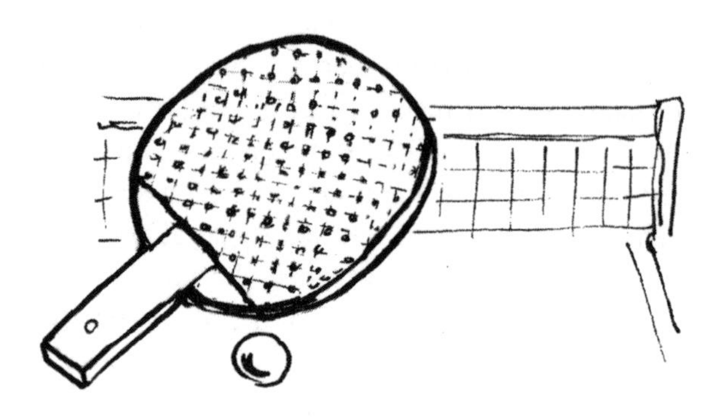

Under the old rules, about 120 points would be scored in a close three-set table tennis match. Each set was of high importance – you could lose one set, but not two. And within each set, the tension was only high during the last six to eight points. Overall, the spectators felt heightened tension for perhaps 20 points out of 120.

Under the new rules, 120 points corresponds to about seven sets. Now a player can afford to lose a couple of sets without feeling a crisis looming, so perhaps just four of these sets will turn out to 'really' matter. But within these four sets, the crisis – 6 to 8 points left – is reached much earlier. Overall, the new rules do seem likely to lead to significantly more pivotal points.

But why stop there? You could make a set first to seven points, in which case a match might have eleven sets! Almost every point in the set would be important, but losing the set would be no big deal.

There is a curiosity that arises out of this. At one extreme, you could have just one set, the winner being the first to score (say) 61 points. At the other, you might have a huge number of sets – 121, say – each of them won by the first player to score *one* point(!) Mathematically, these two extremes are *identical*. In other words, if you take the idea of sets too far, you end up back where you started!

The 'best' balance will be somewhere in the middle, and the modern table tennis rule of best-of-seven sets with 11 points to win a set feels about right. The unwritten rule for tennis, table tennis and the like seems to be that the number of games (or points) to win a set should be bigger than the number of sets to win a match – but not too much bigger.

How rugby changed the points system

Changing the rules of scoring can also have an influence on the style of play. One of the sports that has tinkered with its scoring system several times is rugby union, where the relative values of tries, penalties and drops have changed dramatically over the decades.

	TRY	CONVERSION	PENALTY	DROPGOAL
1891–1892	2	3	3	4
1893–1947	3	2	3	4
1948–1970	3	2	3	3
1971–1991	4	2	3	3
1992–	5	2	3	3

Each change was designed to encourage more positive, exciting play. The most notable change has been the increasing value of the try at the expense of the drop goal. In the very early rules, a drop goal was worth twice the points of a try. Today a try is much more valuable.

The most recent change was to increase the points awarded for a try from 4 to 5. This subtle tweak has had two effects: it means that the average points score in games has increased (good news for players who keep personal tallies in the hope of breaking records); and it also means that a converted try is worth more than two kicks. A team trailing by seven points with only a few minutes to go can no longer hope to level the match with a couple of speculative drop goals, but must instead push for the line.

In theory, that should have made the game more exciting. In practice, perhaps it has only meant that instead of taking a punt at goal, players punt the ball into touch in the hope of a pushover try.

Motor racing points

Formula One motor racing has tweaked its scoring system even more often than rugby, though most of the basic principles of scoring have remained the same for over fifty years. The idea of the winning handful of drivers all getting points while the rest got zero has been around since the 1950s.

Initially, 8, 6, 4, 3 and 2 points were awarded for the first five to finish, and 1 point was also given to the driver who had clocked the fastest lap. This fastest lap bonus was effectively a reward for an added element of thrill in the race. It meant that even a driver who was out of the running had a chance to score right up to the final lap – an intriguing idea that might at first glance appeal to innovators in today's more predictable Formula One races.

The trouble was, the fastest lap bonus point didn't have a positive impact on the race. A team not confident that their car was up to finishing the race, let alone getting into the points positions, now had the temptation of tuning the car at the start so that it was capable of bombing round at top speed for a few laps before breaking down. In other words, the idea of awarding bonus points for the fastest lap backfired (as did the cars, no doubt) and was quietly dropped.

Although there were six races in the 1950 season, only the points from a driver's best *four* races counted. Given the unreliability of cars in those days this seems eminently fair. Otherwise, a couple of unfortunate overheated engines could rob the best driver of a place on the final podium – though arguably the cruel hand of fate is another important ingredient in the excitement of a sporting contest. That year, three drivers with tongue-twistingly similar names took the top three places in the world:

1. Farina (30 points)
2. Fangio (27 points)
3. Fagiola (24 points)

Fagiola came third, but if *all* the races had counted instead of just four, he would have come second; and if the points system being used today had been in operation then, Fagiola would have won the title with 38 points, ahead of Farina's 37 and Fangio's 30. So the scoring system can make a significant difference to the final result.

Three important changes have been made to Formula One in the fifty years since the Famous Fs, all of them aimed at adding to the excitement.

1 The first eight racers now score points rather than the first five (with 10 points for the winner, then 8, 6, 5, 4, 3, 2, 1 for the runners-up), so even followers of the also-ran teams now have something to cheer for.
2 All races count, not just the selected best ones. So a freak accident or breakdown adds a little more drama to the scoring.
3 Most important of all, races that used to last three hours now last about half that time – a reflection of the diminished attention span of the modern spectator, and also of the TV sports producer who wants to pack more action into less time.

There is one other major difference. These days, Formula One is incredibly safe. Nobody says they want to see a horrific crash, but the risk that it *could* happen added a certain frisson that has now been removed. Like the circus without the safety net, the Formula One track with no crash barriers and dodgy cars had a thrill element that even a clever points system will never recapture.

Knockout schemes and World Cup fiascos

One way in which sports build excitement is through knockout tournaments. The great advantage of the knockout is that you can be certain that the winner will be unknown until the very last match. Compare this with cumulative tournaments like Formula One where the top team can build up an unassailable lead, making the final weeks of the tournament an anticlimax.

However, there is no guarantee that the best two teams will make it to the final. Without seeding, even if the two favourites are invincible against all other teams, there is an almost 50–50 chance that they will meet before the final. On one hand this does mean that there is the potential for a romantic clash in the final between a giant and a minnow. On the other hand, nobody enjoys a one-sided match.

For this reason, international tournaments in football, rugby and cricket have all opted for the combined approach of mini-leagues, in which the seeded teams are kept apart, followed by a knockout tournament for those who get through the first stage. All of which is designed to make sure that only the *seriously good* teams make it through to the final stages. Or that's the theory.

Unfortunately, while knockout schemes appear entirely fair and sensible on paper, they can turn out to have unexpected twists that leave the organisers with egg on their faces. Take a look at this elaborate scheme, for example, which was designed to mix lots of exciting matches with a foolproof method of getting the best four teams into the semi-finals. The regulations were these:

- Fourteen countries take part in the tournament, divided into two seeded pools of seven.
- The teams in each pool play each other once, and the top three teams from each pool go through to the next round. (This is just separating out the cream.)

- The six teams in the second round now play another round robin with each other, so that the best four will emerge. However, to save duplication of matches, if two of the teams have already played each other in the first round, they don't play again, the result and points from the first round are simply carried through. (Sounds complicated, but it's just saving time. Every team will end up having played every other team once before the semi-finals.)
- The top four teams now play in semi-finals, with the two winners playing in the final.

What could go wrong?

Lots. This was the format for the 2003 ICC Cricket World Cup. What scuppered it was that two crucial matches were cancelled because countries pulled out for security reasons. Rank outsiders Kenya gained four points when New Zealand cancelled, while a poor Zimbabwe team benefited the same way when England cried off. As a result, two weak teams made it through to the so-called 'Super Six', and to make matters worse, New Zealand qualified too so that the four points Kenya were awarded after the cancellation of their match were carried through to the next round. Despite being trounced by all but one of the strong teams that they played against, the points system and a fluke combination of results between other countries meant that Kenya made it through to the semi-finals, where they were duly thrashed by the favourites Australia. Exciting it certainly wasn't.

In the previous Cricket World Cup of 1999, there had been similar unexpected embarrassment when a loophole appeared in the rules. Teams with the same number of points were ordered according to a complex 'net run rate' calculation. Some smart maths from the Australian captain Steve Waugh led him to realise that if his team beat the West Indies but did so while *batting as slowly as possible*, then both teams could qualify for the Super Six, and Australia would carry through the winning points. The result was a cricket match played in freezing weather where the Australians were deliberately spinning the contest out as long as possible. Never has there been such a stark contrast between what the rules intended and what actually happened on the pitch.

When it comes to controversial qualification in knockout tournaments, soccer is the undisputed champion. In the group stage of the 1982 World Cup, Algeria, West Germany and Austria all beat Chile, and all won one game against each other. But the order of the matches turned out to be vital. The last match was Austria v West Germany. In the event of any tie the teams would be separated by goal difference, and then (if necessary) by goals scored.

The table before the final match was as follows (the top two would qualify):

	PLAYED	WON	LOST	FOR	AGAINST	POINTS
ALGERIA	3	2	1	5	5	4
AUSTRIA	2	2	0	3	0	4
W GERMANY	2	1	1	5	3	2
CHILE	3	0	3	3	8	0

Inspect that carefully: if West Germany failed to beat Austria, then Algeria and Austria would qualify. If West Germany did win, they must qualify, but with which other team? With a narrow victory, Austria would also go through, but with a substantial victory it would be Algeria.

West Germany scored after ten minutes, and very little else happened. Why should it? The score as it stood suited both teams. By playing defensively, the Germans could now protect their lead, while Austria had a cushion of two goals that ensured they would qualify ahead of Algeria.

Shutting the stable door after the horse had bolted, the rules were changed: in group matches, both *final* matches would now be scheduled to take place simultaneously. The intention was that no two teams could go into a match being able to fix the score so that both qualified, irrespective of the other match. But another change partly negated the impact of this move – the tie-breaker was changed from goal difference to the nebulous 'results between the teams involved'.

Fast forward to Euro 2004 when, in the first four matches in a group, Bulgaria lost to both Sweden and Denmark, while Italy drew with both Scandinavian countries, 0–0 and 1–1. Thus, if Sweden and Denmark managed to draw, with both teams scoring at least two goals, both would go through, even if Italy beat Bulgaria 53–0! The reasoning is that, with three teams level on points, having drawn all games against each other, it is who *scores most goals in those drawn games* that is deemed better. So, like Austria and West Germany 22 years earlier, Sweden and Denmark could guarantee each other's progress.

It took until the 89th minute for Sweden to make the score 2–2. For the final few minutes, a frenetic match suddenly became soporific, though nobody seemed to care too much. But suppose that 2–2 score had been reached at half-time. Who could then blame either team for imitating the Austria v West Germany farce?

The real culprits were the organisers, who had failed to consult any mathematicians about the possible consequences of their rules. Had they used goal difference as the tie-break rule, then, if the final group games ran simultaneously, two teams could not begin a match knowing how to collude so as to exclude one of the others. If there is one thing that kills the excitement of a contest stone dead, it is two teams colluding towards a specific result.

How the loser can win in tennis

In lawn tennis, in order to get 'extreme' results, let's suppose the eventual winner wins games or sets as narrowly as possible, but when he loses them, it is as heavily as possible. So he loses games by four points to nil, but wins them 4–2. Now suppose he wins the match in five sets, 0–6, 0–6, 7–6, 7–6, 7–6. In the 30 games he lost, he trails 0–120 in points. He won 18 'ordinary' games, and three tie-breaks, each (say) 7–5.

With the figures given, the loser of the match won 171 points, the winner won 93 – so the loser won 65 per cent of the points.

15

THE PERCENTAGE GAME

Play safe or take a gamble?

There are certain clichés that spread like viruses from one sport to the next, and one of the most popular is the notion of a 'percentage game'.

Bernhard Langer may be heard talking about playing 'percentage golf' up the fairway. The Wimbledon football team of the 1980s were often accused of playing 'percentage football'. You will find 'percentage plays' in snooker, tennis, baseball, rugby, basketball and no doubt in every other sport where the participants have a choice of tactics. But what, if anything, does this phrase mean?

'Percentage play' generally has negative connotations. It is associated with safe, calculated but rather dull play, lacking in adventure or innovation. Percentage players may be clinical, but they don't pull in the crowds.

The fundamental idea of percentage play must be to improve a player's or a team's chance of achieving their object. By adopting Tactic A (e.g. 'always shoot when within ten yards of the penalty area'), they might expect to win 30 per cent of the time; by adopting Tactic B (e.g. 'don't shoot until you're inside the penalty area') they might expect to win 60 per cent of the time. Tactic B would then be the 'percentage play', because it has a higher percentage success rate.

But if it were this simple, then the cliché of percentage play would be vacuous. Any team or player that *wasn't* indulging in percentage play would be letting down itself and its supporters.

The true meaning of the phrase is surely more subtle, so let's delve further into a number of different sporting situations. What we'll find is that a true 'percentage game' is not always the most obvious, and it can sometimes involve more risk than the other options. It all depends on the particular players and on the circumstances of the match at the time.

Above all, it depends on the *ultimate objective*: a snooker player may take on a risky shot that keeps him on course for a 'maximum' break of 147, and an associated cash prize of £147,000. If his priority is money, he should take the risk: but if the ultimate object is to win the match, he should go for the safer shot and abandon the 147 glory. Always know what you are trying to achieve.

We've discussed elsewhere a number of situations that might be described as percentage play. In darts, for example, aiming at triple 19 or triple 16 will deliver a higher average points score for most players than aiming at triple 20. But playing safe in the short term doesn't necessarily deliver the best result in the longer term. A slower tennis serve is less likely to be a fault than a fast serve, but we saw in Chapter 8 that attempting a fast serve first, followed by a slow serve, is more likely to win the point. So players who cop out of doing something that risks failure are sometimes merely delaying the inevitable.

And sometimes 'risky' moves hold no risk at all. A long snooker pot that might only have a 20 per cent chance of going in is often still worth taking on if the balls are likely to end up in safe positions should the shot be missed. Commentators usually refer to this situation as a 'shot to nothing'.

Golf: flamboyance versus consistency

One sport where risk and caution are forever battling within the player's mind is golf. In particular, is it better to be a steady player with no frills,

as Nick Faldo was in his prime, or to be a flamboyant player known for moments of inspiration and calamity, like Seve Ballesteros?

It isn't often that great mathematicians involve themselves with theorems about sport, but one exception is G H Hardy. Hardy was an academic who delighted in the fact that much of pure maths had no direct application. However, he spent a great deal of his leisure time thinking about sport, and this led him to develop a golf 'theorem' which dealt with the issue of consistency versus flamboyance. Hardy concluded that steady golfers will tend to beat erratic ones of similar ability. His analysis was quite complex, but it is possible to summarise his argument in a more descriptive if less rigorous way.

Let's suppose that a player only has three types of shot:

Good (G)
Ordinary (O)
Bad (B)

And for the sake of making the analysis easy, let's also suppose that each 'Good' shot that a player makes contributes to him *gaining a stroke*, and each 'Bad' shot means he *loses a stroke* (this is equivalent to saying a Bad shot is a shot that doesn't move the ball at all). An 'Ordinary' shot means he is on course for par. If a golfer is extremely consistent, then on a par four hole, he would hit four Ordinary shots (O O O O) then end up with par.

Let's now turn our attention to the inconsistent player, who for ease of argument we'll say is equally likely to hit a Good or a Bad shot. When he plays a par four hole, there are many possible outcomes, for example:

G O O (Good, Ordinary, Ordinary) resulting in a birdie
O O G (also a birdie)
G O B O (par)
B O G B O (this is a bogey)

You can have fun playing your own imaginary golf holes using this terminology, but bear in mind that the final shot can't be Bad (by definition a bad shot must miss the hole). So, BOOBOO is possible for a double bogey, but OOBOOB is not.

Now for Hardy's reasoning. On a par four hole, an erratic golfer needs at least one Good shot in his first *three strokes* if he is to get a birdie. (This should be obvious with a moment's thought.) However, it is possible for him to get a bogey even if he only has one Bad shot in his first *four strokes.*

Since we're assuming that Good and Bad shots are equally likely, the chance of a Bad shot in the first four strokes must be higher than the chance of a Good shot in the first three. (It's the same as the more familiar argument that you are more likely to get at least one head in four tosses of a coin than in three tosses.)

Of course, a full analysis must take all possible stroke sequences into account, and tot up the overall chances of 2, 3, 4, 5, etc., strokes for a hole. But the conclusion indicated by this argument does indeed follow. In short, according to Hardy's model, an erratic golfer will have a slight bias towards bogeys rather than birdies.

In the case of golf, therefore, it looks like steady but unglamorous play will tend to win the day. And it can certainly happen. In the 1987 British Open, Nick Faldo famously won after making par on every hole in the final round.

However, while this argument for 'percentage play' might be true *in terms of average scores*, an erratic golfer might actually expect to win more *prize money* overall. This is because prize money is heavily weighted towards the top finishers. A player who wins one tournament and comes 50th twenty times will earn far more prize money than someone who ends 20th in every tournament, even though his average score is worse. Normally, a preponderance of pars is not enough to win a golf tournament. What's needed is a flash of brilliance on the day (combined with a bit of luck).

Nothing to lose

In many team sports, there is potential for percentage play in the form of the tactical substitution. The most clear-cut example is in football. Suppose the City coach has a choice between two players:

- Mal: a good ball winner, but with little attacking capability
- Steve: he can't tackle to save his life, but he does score goals.

In a League match, City are away to Rovers, a strong home side. A draw would be a good result, and that is quite possible with Mal in the team; if Steve plays the whole game, defeat is very likely. The coach correctly decides to start with Mal, and will only think about a change if they fall behind.

But if Rovers do take the lead, should a substitution be made, and if so, when? Operations research experts have analysed this conundrum and offered their solution. Although Mal is better for the team when they are ahead or level, Steve is better when they trail. Why? Because if City are 1–0 down, their chances of scoring and therefore salvaging a point are slim with Mal in the team, whereas with Steve they are higher. *But Steve should not come on too early*. Even if City go one down after ten minutes, it may be best to keep Mal on for another hour before swapping him for Steve. For if Steve has to play for most of the game, he might work his magic once to get the equaliser, but his defensive weakness over a long period would be too high a liability.

This tactic of raising the stakes towards the end of a match is the 'nothing to lose' philosophy. As the old saying goes, you may as well be hung for a sheep as a lamb. It arises in plenty of different situations:

- In ice hockey, a team that is trailing with not long to go will often adopt the tactic of pulling the goalie and bringing on an attacker instead. Just as in the football example earlier, the only chance of the trailing team getting anything from the game is by scoring, and even if conceding a goal is *even more likely* than scoring once the goalie has been pulled, it can be worthwhile taking the risk. Moreover, unlike football where reverse substitutions are not permitted, if the gamble works, the goalie can be restored to his position;

- A long jumper who needs to achieve an extra 10 centimetres if she is to qualify for the next round and only has one jump to go might reasonably take the risk of trying to set off much closer to the 'foul' line than normal. It is more likely that her shoe will cut into the plasticine, but the risk will be worth it if she has a chance of taking off a few centimetres further ahead than she did with her earlier clean jumps.

In fact this long-jump example is the reverse of the Fast–Slow tactic for tennis serves we described in Chapter 8. You may recall that in that chapter, we showed how the tactic of Fast–Slow for a tennis server is always superior to Slow–Fast. For 'fast' read 'risky' and for 'slow' read 'cautious'.

In long jumping, Cautious–Risky can be the superior tactic. The reason is that the nature of the opposition is different in the long jump from that in tennis. In tennis, you serve directly to one opponent, and your object is to win that particular point. A long jumper is competing against perhaps a dozen opponents who will all be using their own strategies to achieve one splendid jump. It may well make sense to start with a cautious jump that puts some decent mark on the board, and subsequently attempt a series of jumps with the risky strategy, in the hope that one of them is good and extends this mark.*

Try your hand at some percentage play

Imagine you have agreed to play a game of darts against an opponent who plays at the same standard as you. There will be an unusual rule for deciding the outcome of this contest: the winner of the match is the first to win two legs in a row (the winner of a leg is the first to get to 501 points). As is normal in darts, you and your opponent will take turns to throw first in each leg.

As we mentioned in Chapter 7, the player who throws first in a game has a better chance of winning it, especially if both players are good. You now win the toss. Do you . . .

(a) Choose to throw first in the first leg?
(b) Ask your opponent to throw first in the first leg?
(c) It doesn't matter; it makes no difference to the probability of winning.

Answer at the end of the chapter.

Percentages and the horses

A chapter on risk and gambling would not be complete without mention of the sort of risk-taking that is done in the bookies' tent. When sports fans gamble money, they are playing exactly the same percentage game as the people they are following in the arena. One huge difference is that the bookmaker's odds are written down in black and white, whereas the sports player can only follow a hunch as to what the real odds of a particular strategy will be.

'Fair' odds should be a genuine reflection of the chance of a particular outcome. If Lucky Lady is 9 to 1 against to win the Cheltenham Gold Cup, this would be a completely fair bet if the horse has a 1 in 10 chance of winning. But there are two crucial factors:

* In the 1968 Olympics, Bob Beamon shattered the world record with his first jump, hitting the take-off board in the ideal spot. So his first jump was definitely 'Risky'. But had he fouled on that jump, with only two more attempts to put some mark on the scoreboard, we suggest that his next attempt would have called for Caution.

- The bookmakers need to make a profit, so the odds are 10 to 20 per cent shorter than they should be (with odds of 9 to 1, Lucky Lady's real winning chance might be 1 in 12, rather than 1 in 10);
- Even more important, the odds are heavily influenced by the psychology of the other punters. It's not the chance of Lucky Lady winning that matters, it's what the betting public *thinks* the chance would be. The more punters bet on a horse, the shorter the odds become.

The golden rule in all sports betting is that the only sure winner is the bookmaker. The canny bookie redistributes about 80 or 90 per cent of stakes to that race's 'winners', and keeps the rest. In the long run, it's then a matter of your personal skill, knowledge and good luck as to whether you end up losing just a little or losing a lot. There are, however, some tactics that do statistically seem to move the odds a little more in your favour.

In the glamorous races, romantic notions tend to influence a lot of the punters, which in turn distorts the odds. Horses with popular jockeys or with heroic stories tend to attract more bets than they should, especially in races like the Grand National. For the professional gambler, those horses therefore represent a bad bet, since the odds on them are shorter than they should be.

In the same way, patriotic fervour means there will be a tendency for people to bet more on their own country than on other countries, and in order to balance the books, the bookies shorten the odds to counteract this. So if you want to back England to beat Spain at soccer, bet with a Spanish bookmaker.

Another thing all professional gamblers know is that outsiders are generally worse 'value for money' than favourites. If the odds offered for a horse are 50 to 1, it is very likely that the fair odds are more like 250 to 1. Backing such rank outsiders tends to lose 80 per cent of your money, while backing well-favoured horses might lose only 10 per cent, on average.

But if you spend an afternoon at the races, there's a rule of thumb about when to back an outsider. At the start of the afternoon, punters have lots of money in their wallets, and may therefore be inclined to back the favourites, to enjoy the feeling of winning something. At the end of the afternoon, however, the majority are bound to find themselves in the red. The only way for them to end with a profit is by betting on a horse with long odds and hoping that it wins. There is therefore a tendency to bet on the outsiders in the last race of the day.

If there is truth in this theory, then as a percentage player it gives you a sound basis for your own strategy: do the complete opposite of what the majority are doing. In other words, bet on the outsiders in the early races, and go for the favourite in the last race.

But if you still end up losing, don't blame us. We didn't choose your horse.

Answer to the percentage play puzzle

You might think that by throwing first in the first leg, you have an immediate advantage, because you are more likely to go 1–0 up. But according to the probability calculations, the right answer is (b): to maximise your chance of winning the overall match under these rules, you should ask your opponent to throw first in the first game.

Here is an informal explanation of why you should give your opponent the advantage in the first game. Since the only way to win this contest is to win two consecutive legs, you know that you are going to have to win one leg where your opponent is throwing first. Therefore, it pays to have as many opportunities as possible to play the opponent when he's strong, since the more opportunities you have, the more likely it is that you will win against the odds in one of the games.

However, this result is extremely counter-intuitive, and not easy to justify without some probability calculations. You'll find these in the Appendix.

16

HOW TO WIN GOLD

And how many medals your country might expect

Ethelbert Talbot is not a name that naturally trips off the tongue when it comes to great sporting quotations. It was he, however, who in 1908 delivered a sermon at St Paul's Cathedral that included what was to become an immortal line:

> *The important thing in these Olympic Games is not so much winning as taking part.*

Pierre de Coubertin, founder of the modern Olympics, heard the sermon and adapted this line to become the creed for the movement. It's a creed that sits comfortably with the general public, but not necessarily with the media or with government, who deep down know that from their point of view, it's not so much the taking part that is important, but the tally of medals.

For years, the greatest battle to top the medals table was between the USA and the former Soviet Union. For the rest of the twenty-first century, it may well become China versus the Rest of the World. But it's not just the big countries who want medals. The tiny island nations cherish the

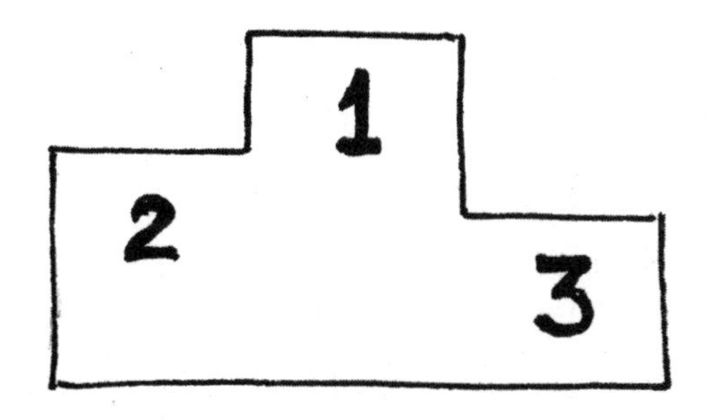

achievement of a single bronze just as much. The question is, are there tactics that a country can use in order to get more medals? And when does the athlete of modest abilities have the best chance of getting onto the podium?

Reducing the competition

There's a simple principle for improving your chances of winning gold. Keep the number of competitors down to a minimum. The fewer rivals you have, the better your chances of winning.

One of the best ways of excluding other competitors is by pricing them out of the market. So for a country in search of gold, the 100 metres race is unlikely to be fertile ground, the problem being that *everyone* can take part because the cost of the equipment and the technical facilities needed are small. On the other hand, if the sport in question needs, say, an expensive yacht built to the highest technical specifications, then half the countries in the world are excluded simply on cost grounds, and another considerable number will be excluded because they are landlocked, and don't have appropriate facilities to practise on. We don't expect a Nepalese yachtsman to be standing on the podium collecting a medal any time soon.

Sometimes, intrepid teams can achieve remarkable things despite the lack of amenities – the feats of Jamaica's bobsleigh teams in the winter Olympics being a glorious example. But these are the exception.

Other sports requiring extremely pricey equipment include cycling, rowing and (if you can call a horse 'equipment') show jumping. Competitors in these sports would rightly point out that the competition for medals is still extremely fierce, but the entry-level costs mean that a huge proportion of the world's population is effectively excluded from taking part.

As we pointed out in Chapter 9, there are also some sports where an individual has the opportunity to win several medals using the same basic skill – in swimming for example. So the cynical country looking to improve its medal count might hit the jackpot by concentrating on exclusive sports where several medals are available. In the eyes of the world looking at the medals table, all medals are equal. But to misquote George Orwell, some medals are more equal than others.

The benefits of size

The link between money and medals has been explored by economists on a much larger scale. There turns out to be some correlation between the wealth of a country and the number of medals that it wins. There are two reasons for this. A country with a high Gross Domestic Product (GDP) probably has a large number of people, and the higher the population, the more likely it is that there will be a star athlete somewhere within it. But wealth also brings with it the luxury of being able to invest in sports amenities, and big countries like to make their presence felt, too, so it becomes politically important to register on the medals table.

Calculating the medal points using the traditional method of 3 points for gold, 2 for silver and 1 for bronze, you can produce a scatter chart of medal points against GDP. Many countries do seem to fit quite closely to the dotted line that we have somewhat arbitrarily drawn through zero and through the UK's point on the chart. Those countries that are above the dotted line have performed better than the link would suggest, while those below have performed worse.

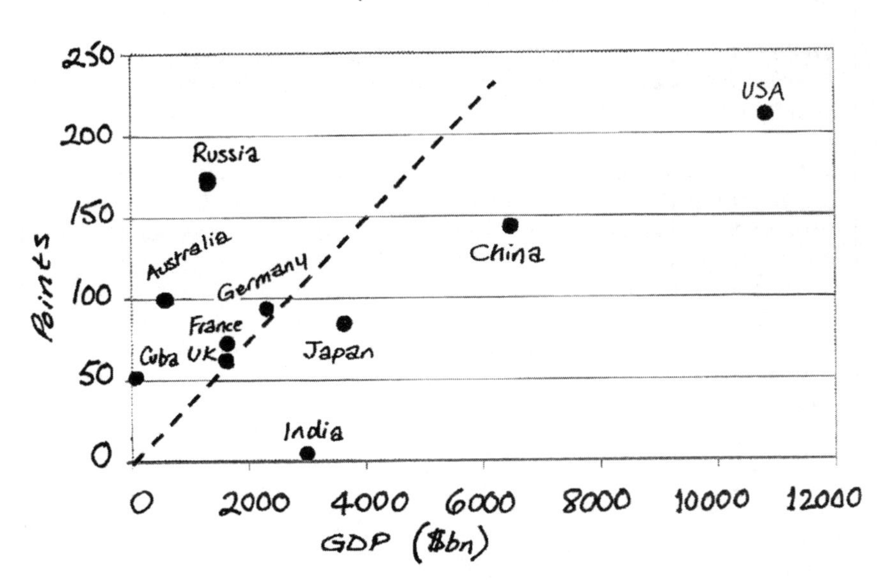

Data from 2004

The diagram highlights some notable exceptions, those countries close to one axis or the other. Of the leading countries, Australia performs outstandingly. Its passion for sporting excellence is reflected in far more medals than a country of its size and wealth might reasonably expect. Cuba too punches well above its weight (literally, since boxing medals have traditionally made up the majority of its points). Russia also does well on this table – it has a low GDP for its large population!

On the other axis, the country that stands out by a mile is India. It seems incredible that a country of over a billion people struggles to get even a single medal. Part of the explanation is down to the country's overwhelming passion for cricket, a sport that hasn't appeared in the Olympics since 1900. That's another rule for getting medals: make sure you concentrate on sports that actually feature in the tournament.

Why big countries win the relays

There is another, more subtle reason why large countries might be favoured when it comes to winning medals. Many medals are for teams, rather than individuals. This doesn't just mean the traditional team sports like hockey and basketball, but sports where individual performances are added together, such as gymnastics, and the relay races in athletics and swimming.

Team sports favour big countries. You can just about imagine a Liechtenstein runner winning the 400 metres, but you can't imagine that country winning the 4 x 400 metres relay.

Indeed, as the number in a team increases, there is a mathematical argument that the chance of a small country winning medals decreases exponentially. In a sport with N members in a team, the number of medals a country might expect should be proportional to x^N, where x is its fraction of world resources. For example, if country A has twice the resources of country B, it might expect twice as many individual medals (N = 1), but $2^4 = 16$ times as many medals in team-of-four relays (or coxless fours at rowing). A description of our reasoning for this claim can be found in the Appendix.

Accidents will happen

If you are only moderately good at your sport, it will also help your chances of winning a medal if the sport you are competing in is one where accidents happen, especially if the accidents have a disproportionate impact on the end result. This is the snakes-and-ladders principle. If you are a skilled games player you are most likely to demonstrate your prowess in a high-skill, low-accident game like chess. The greatest leveller, on the other hand, is snakes and ladders, where skill is ruled out altogether.

Accidents happen in all sorts of sports, but in some cases they are not catastrophic. A pole vaulter whose pole snaps as she's about to go over the bar does at least have another go. But a show jumper whose horse suddenly gets the yips can move from gold medal place to also-ran in the matter of just three jumps.

Even more catastrophic are some of the treacherous races. In 2002 in Salt Lake City, Australian speed skater Steven Bradbury was trailing in last place in the final, when suddenly ahead of him there was an almighty clatter as the top skaters clashed and tripped each other, with less than half a lap to go. Bradbury cruised through the pile of bodies to claim Australia's first-ever Winter Olympics gold medal. It was a clear case of the top players taking each other out, so that the outsider could come through and scoop the prize.

Ice can often be a good leveller because it is, well, slippery. Although this hazard should affect all competitors equally, in some situations this is not the case. Take the example of downhill skiing. The more compacted snow becomes, the faster the surface gets. This is generally a good thing for skiers, though there is a down side, in that if the weather conditions change then icy snow can actually become *too* slippery as a tournament progresses.

In an attempt to remove any advantage that this might give to some skiers over others, the normal practice is for the skiers to perform their first downhill race against the clock in one order, and then to do the second in reverse order. If the conditions do vary, then all competitors should get a combination of good and bad. That's fine in theory, but in practice this will only be completely fair if the change in conditions is *linear*.

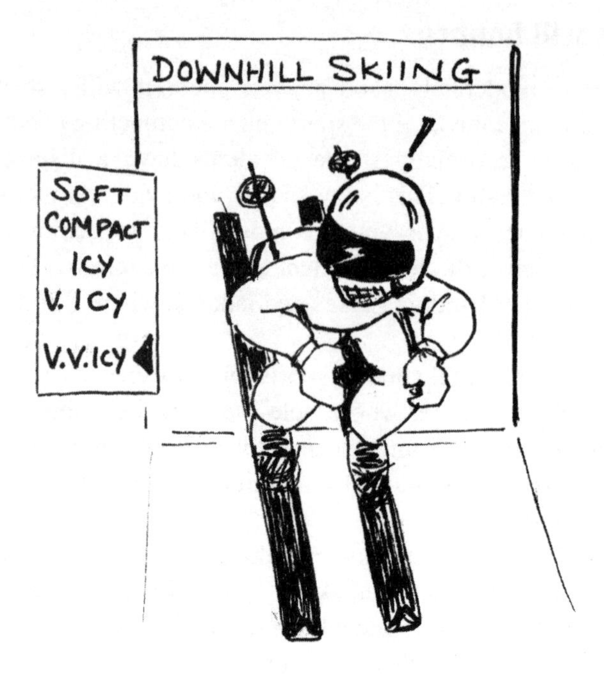

So, if the speed of descent of a skier increases, so that each competitor gains 0.02 seconds over the preceding one, the combination of two downhill runs should be the same for all competitors.

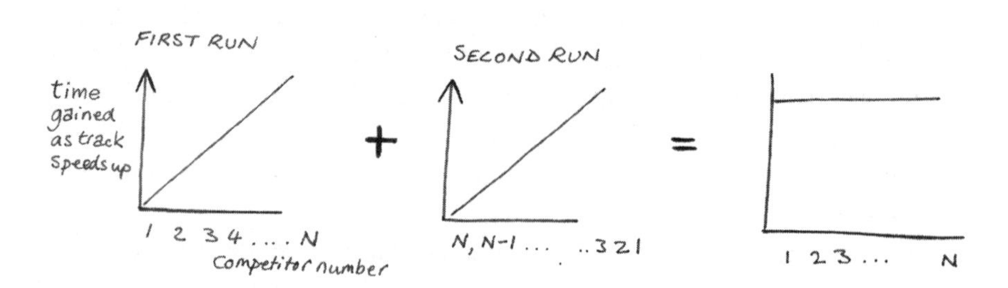

But in most real competitions, the change in conditions is not linear. Typically, the first few skiers will do most of the compacting of the snow, so the most significant change in conditions happens early on. And as soon as the graph of time gained against number of skiers becomes a curve, or even if the slope of the graph changes from one round to the next, then some skiers will be favoured over others when the two graphs are combined. The most favourable position in the draw will depend on how the surface changes, but the fact that luck enters it at all is an advantage to the country that is struggling for medals.

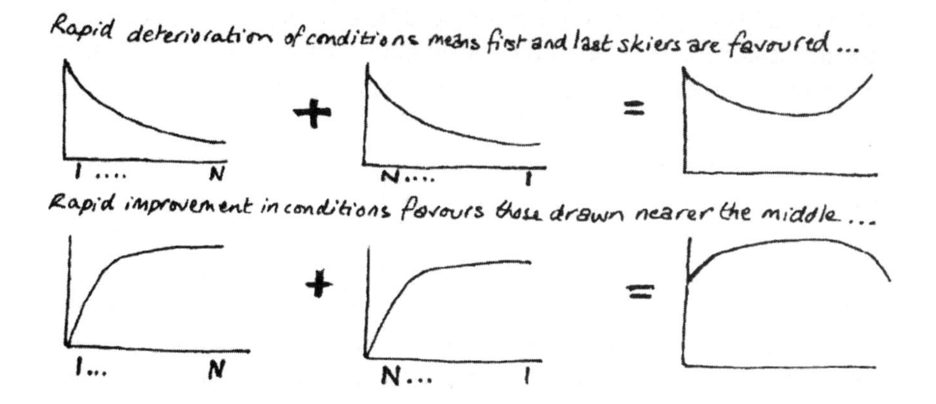

The luck factor

Luck can manifest itself in other ways, too. In those sports where there is a knockout elimination, a weak performer might well avoid strong opponents, simply because another competitor has already knocked them out in an earlier round. In such tournaments where the performers are not seeded, it is theoretically possible for all the top players or teams to find themselves in one half of the draw, leaving the way clear for the minnows in the other half to get through to the final. But the arrangements that exist in unseeded, random knockout competitions like the FA Cup don't generally apply in major medal tournaments. So the chance of a minnow making the silver medal play-off in badminton, table tennis or archery is extremely low.

Luck will also play a more significant role if there are not enough contests to ensure that the top performers are guaranteed to shine through. To take an extreme example, if rifle shooting involved one chance to shoot a single bullet at a target, with the closest to the bull winning gold, then this would present a great opportunity for a mediocre performer to be elevated to medal status. However, neither rifle shooting nor any of the other Olympic sports seems to offer such rich potential for lucky medals.

But if golf happens to make it to Olympic status, as many are hoping, then whole new opportunities will be available. The fact is that the best player in the field usually *doesn't* win a golf tournament, even one of the four majors. The reason is that nearly all the 150 or so players who enter the tournament are very, very good, and even 72 holes isn't nearly enough to eliminate the effects of luck. A single poor shot can cost a golfer two strokes, and two or three of these, or a couple of flukey holes in one, can turn the leaderboard upside down. Counting all four rounds does go a substantial way towards reducing the role of chance, but by no means the whole way.

If you rank the players according to how well they do compared with par, on average over a couple of years, someone – Tiger Woods, Ernie Els, Vijay Singh – will come out on top. But the most likely number of majors that any one of these players will win in a single year is zero. The chance that any of the world's top ranked players would win Olympic gold is probably lower than would be the case in any other sport. This definitely makes it a sport for the aspiring Liechtenstein medallist to think about.

The threat of drugs

There is, regrettably, one other way in which countries or individuals attempt to increase their tally of medals. Drugs. The extreme cases of drug taking that distorted the results in the 1960s and 1970s may no longer occur, but the frequency with which top athletes are found to be over the tolerated limits shows that the problem has not gone away.

There is a constant battle between the drug testers and those who seek to disguise the drugs they have taken. In an ideal world, drug testing would be a precise and 100 per cent accurate science, but unfortunately this is not the case. And the consequences can be extremely harsh.

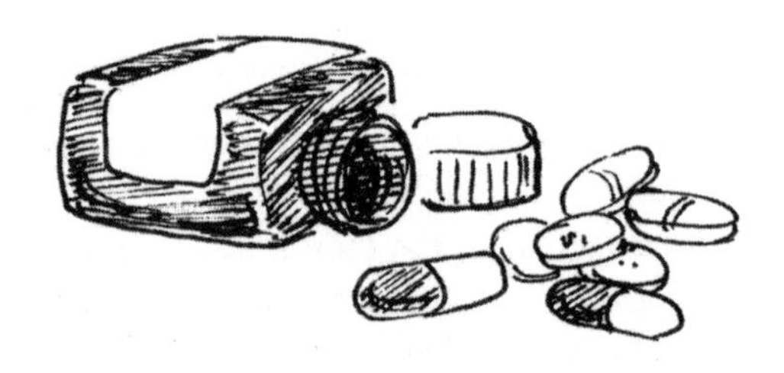

The American middle distance runner, Mary Slaney, was found guilty of illegally using testosterone, and banned from the Atlanta Olympics. Athletes are at their peak for a very short time in their lives, and if they are *unjustly* accused of drug abuse, they may suffer significant financial loss, as well as losing the chance of Olympic glory for ever.

They are seldom convicted of drug use on the *direct* evidence of being seen injecting themselves or consuming banned substances. It is usually by the indirect process of making a chemical analysis of a sample that they provide. But when the substance being investigated is produced naturally by the body, as well as artificially via banned drugs, this approach has clear problems.

Slaney's drug test relied on the so-called T/E ratio, a comparison of the relative amounts of two substances normally found in urine samples; usually, these two substances have very similar concentrations, so the T/E ratio should be close to 1:1. The authorities set the threshold that a T/E ratio of more than 6:1 was sufficient to provide 'proof' of drug use. Slaney's T/E ratios on two samples in 1996 were roughly 10:1 and 12:1, and so she was banned.

But was that threshold reasonable? Any test must address two problems: it must be *sensitive* – a high proportion of drug users should fail the test; it must also be *specific* – a very high proportion of innocents must pass the test. But T/E values for men and women are different, and also vary naturally with age, ethnicity, diet, lifestyle, etc., and it is possible for the test to give an incorrect result.

Let's try some numbers. Suppose we have 1,000 athletes, a group which happens to include 100 drug users. To try to protect the innocent, make the specificity 99 per cent, that is, 99 out of 100 innocent athletes pass the test. Let's also suppose that the sensitivity is perfect, so that if an athlete has taken the drug, he or she will certainly be found out when tested. This seems like an extremely fair and rigorous set of conditions. But the consequences could be alarming.

If the specificity is 99 per cent, this means that 1 per cent of the innocents – 9 out of the 900 athletes – and all 100 of the users will fail the test. So out of 109 who fail the test, 9 are clean. In round figures, 8 per cent of those banned for failing this test are completely innocent. Such an error rate would be unsustainable. So even a drug test specificity of 99 per cent is too low!

In the USA, about 90,000 urine samples are tested annually. Suppose the specificity could be raised to 99.9 per cent. If *all* these athletes were innocent, then we would still get 90 positive tests. And, of course, raising the threshold means that the sensitivity will drop – more drug users will get away with it.

There is no escape from this inexorable logic: the more extensive a programme of random testing, the higher the threshold needs to be set to ensure innocent athletes are not convicted. It has been calculated that, on present numbers of tests, to reduce the chance of *any* innocents being wrongly declared users to below 1 per cent, then the testing capabilities of the entire world would be needed even to set the threshold!

We don't know if athletes like Slaney were innocent, but the fact is that whatever threshold is set, there will be *some* innocent athletes who fail the test and some guilty ones who will pass it. Finding the right balance is not easy. And so long as athletes need to be given the benefit of the doubt, this back door to the medals table will unfortunately remain open.

17

EXTRA TIME AND THE PENALTY SHOOT-OUT

How to end the game

It's time to blow the final whistle, pull up the stumps and wave the chequered flag. But there's a problem. We still don't know who the winner is.

In many sporting contests, a draw is not satisfactory. So, if the contestants are tied when the match has run its full course, there needs to be a way of resolving who won.

Some competitions have ways of deciding a winner when the leaders are tied. In long and triple jumps, ties are broken by looking at whose second-best jump was longer. But this notion of looking back lacks the dramatic punch of carrying on until one side does something positive to win the game. Most sports opt to play on, often in the form of sudden death. This applies, for example, in snooker (where the black is respotted), golf (the players go back to a nearby hole) and baseball (extra

Sudden death in bowls

innings are played until one team scores more than the other). It even happens in bowls, with the winner of the next 'end' winning the match.

Extra time doesn't always guarantee a positive result. In 1939, the normal five-day limit of a Test match was waived in a so-called 'Timeless Test' to decide the series between England and South Africa. Incredibly, after nine days of play, the game still hadn't been resolved, and the English cricket team had to abandon their run chase in order to catch the boat home.

So on rare occasions, you have to put up with a draw.

Extra-time options

But when most people think of 'extra time' they think of football. The low-scoring nature of football means that draws are common, and hence when the elimination of one team is necessary, extra time is common too. Unfortunately, even an extension of 30 minutes isn't certain to resolve the game. Three extra-time options have been tried over the years. They are:

- play on for 30 more minutes
- Golden Goal (first team to score in time added on wins the match)
- Silver Goal (team that is ahead after 15 minutes of extra time wins, otherwise play another 15 minutes)

Which is the best? It will be the one that delivers excitement and a fair outcome quickly. One way to resolve this question is to make some calculations.

Suppose teams play normally during extra time. Assuming that a game of 90 minutes will feature just under three goals on average, the average number of goals in 30 minutes of extra time will be about one. There will be a clear result whenever there is an odd number of goals, or when there is an even number that don't split equally. Assuming goals occur randomly as we discussed in Chapter 10, this chance works out at just over 50 per cent. So we might expect half of all soccer matches that go to conventional extra time to be settled during that period. Those aren't great odds for TV schedulers who want the game over with as quickly as possible.

The chance that a 'Golden Goal' will determine the result during extra time is simply the chance of any goal being scored within 30 minutes. If players played the same way as they do normally, the chance of the match being decided by a Golden Goal would be about 60 per cent, i.e. rather higher than a result in ordinary extra time, as you would expect. But is this what would happen in practice? There is evidence that the Golden Goal rule significantly alters how teams play – you can't afford to let a goal in, so tactics will tend to become more defensive. If this *halves* the scoring rate, then the chance of a decision drops to about 40 per cent. For a Golden Goal to be more likely to settle the match than conventional extra time, the average scoring rate must be at least 80 per cent of the normal rate. One reason why Golden Goals have fallen out of favour could well be because they failed to get close to this threshold level.

And what about the 'Silver Goal'? If teams play normally, and otherwise using the same assumptions as before, it turns out that the Silver Goal system should decide a winner nearly 60 per cent of the time. In reality there might be a tendency for more caution, though not as much as with Golden Goal. This will reduce the chance of a decisive result. So while Silver Goal has some attraction as a compromise solution, traditional extra time may be the best of the lot.

The order of the shoot-out

If extra time fails to separate the teams, matches are now settled on penalties. But who should take them, and in what order? The candidates for taking penalties are all the players, including the goalkeepers, who are still on the pitch at the end of extra time. Five of them must be nominated to take the initial responsibility.

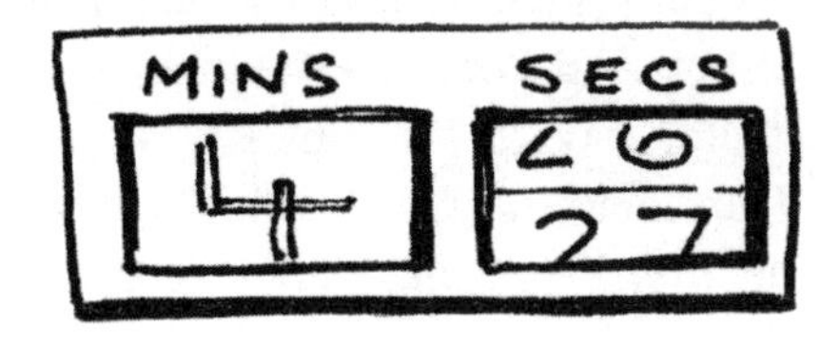

It is plain that you should first choose the five players whose scoring chances are greatest. But once this is done, mathematically it doesn't make the slightest difference what order they are sent out in and so, for reasons we will explain in a moment, you may as well put them in order of ability, best first. The chance of your team winning the shoot-out after five kicks is the same.

This blunt claim surprises some people, but it follows from very straightforward calculations, and one key assumption: *that each player's scoring chance is unaffected by what others have done.* Without that assumption, the order can make a difference to the likely outcome, but in subtle ways, depending on how players react to pressure. One piece of research suggested that teams actually did better in the penalty shoot-out if the weakest of the five kicked first. If this really is the case, then it is because the probability of scoring for each player changed due to psychological factors.

If the score is level after the first five penalties, there is a sudden-death format: a series of head-to-heads with new players, until one team scores and the other misses. At this stage, the best approach is the common-sense one: select the remaining eligible player with the best chance of scoring.

It is conceivable, though very unlikely, that the scores will still be level after every player has taken one penalty. The rules insist that sudden death continues, but the order of the players is the same as before. And now, you will regret it if you did not put the first five players in descending order of their scoring chances. So think ahead. Put the eleven players in descending order of their likelihood of scoring, and send them out in that order. Not only is it the simplest, it also has logic to commend it.

From the fan's point of view, there is another critical question. What difference does scoring or missing the first penalty in a shoot-out make to the eventual outcome? If your team miss the first one, do you have serious room for optimism, or should you leave now and expect the worst?

The exact answer depends on how likely players are to score anyway, but we can get a good feel for the importance of that first penalty if we assume that all the players, on both sides, have exactly the same chance – call it p – of scoring with any shot. At the outset, either team is equally likely to win.

If p were quite small – say 25 per cent – it wouldn't be a disaster to miss the first kick. The chance of winning would still exceed 40 per cent; and scoring with the first kick here would be excellent – already the winning chance is above 70 per cent. But even an amateur would expect to do better than 25 per cent.

At the other end of the scale, when p is about 85 per cent, missing with the first penalty is very bad news, because you are now relying on the opposition missing more than you, and that is unlikely to happen. In this situation, missing the first penalty reduces the overall chance of winning to below one in four. If the first kick scores, as is expected, the chance of winning increases, but only from 50 to 55 per cent.

For most games, however, we reckon that 75 per cent is a reasonable level for p, and if under that assumption you miss the first kick, your overall winning chance has dropped below 30 per cent. Scoring takes that chance up to 57 per cent. So if you reckon a 30 per cent chance is bad odds, then you have reason to feel pessimistic if your team takes the first penalty and misses.

As the chance of scoring with any kick ranges from an absurdly low 15 per cent to an optimistic 85 per cent, the *difference* in your overall winning chances, according to whether the first kick scores or misses, is around 25 to 33 per cent. So that figure could be regarded as a measure of the *importance* of that first kick.

Penalty shoot-outs are widely seen as a lottery, but they are far better than the method used in the 1968 European Championship. There, Italy and the Soviet Union drew their semi-final 0–0, after extra time, and Italy went on to 'win' the match on the toss of a coin. It's hard to imagine the public could tolerate such a method these days. For a start, there would be nobody to blame.

The final whistle

We have finished almost where we started, with the maths of penalties, where some mathematical thinking in the manager's toolkit could make the difference between victory and defeat.

Not everyone, however, sees it that way. Some years ago, an Australian student wrote to a cricket administrator, asking for figures so that he could investigate the possible effects of varying the traditional batting order. The idea that maths might possibly be of interest to a sportsman was too much for the administrator: 'Any analysis that you suggest would be of no value . . . greeted with scorn by those with a proper understanding of cricket . . . An attempt to reverse the inherent unpredictability of the game . . . is not to be encouraged.'

Some minds are closed to the notion that maths and sport could, or even should, ever interact. But for those who see it differently, we hope that far from detracting from the charm and unpredictability of sport, this book has in some small way added to it further.

APPENDIX

For those who wish to see some detail of the maths that would have been distracting in the text – or might even have stopped you in your tracks – we collect it together here, in chapter order.

In **Chapter 2**, there was a table showing the chances 'Beckham' might score from a penalty, according to what he and the goalkeeper did.

Goalkeeper's choice

Beckham's choice		Stand still	Dive to one corner
	Aim straight	30%	90%
	Aim at a corner	80%	50%

The key point is that if Beckham aims straight, the keeper is better off standing still, whereas if he aims at a corner, the keeper's best tactic is to dive. Also, if the keeper stands still, Beckham should aim at a corner; if he dives Beckham should aim straight. Neither player has what is called a *dominant strategy* – one that is better than the other, no matter what the opponent does.

As the chapter pointed out, game theory says that both players should mix their strategies at random, in Beckham's case aiming for the corner two-thirds of the time, while the goalkeeper should dive five-ninths of the time. These figures are derived by finding the ratio where the chance of scoring (or saving) is the same, whichever of his two tactics the other person uses.

We can check that the 1 to 2 ratio does indeed deliver the same chance of Beckham scoring, regardless of what the goalkeeper does. For example, suppose the goalkeeper opts to stand still. To find Beckham's scoring chance look down the first column. This chance is:

$$\frac{1}{3} \text{ x } 30\% + \frac{2}{3} \text{ x } 80\% = 63.3\%$$

the figure we quoted in Chapter 2. If, instead, the goalkeeper opts to dive, the computation using the second column gives the same final answer:

$$\frac{1}{3} \text{ x } 90\% + \frac{2}{3} \text{ x } 50\% = 63.3\%$$

This game theory approach has the advantage that the goalie can dance around and bluff as much as he wants, but Beckham can just ignore him and do exactly what he planned to do anyway with a guaranteed success rate of 63.3 per cent. Of course, if the goalie gives away real information about what he is going to do, Beckham can adjust his tactics and increase his chance of scoring.

* * *

In **Chapter 4**, we noted that it might happen that 'rounding' could lead to injustice. Here is an illustration from gymnastics. In the vault competition, the competitor is given the average score from two vaults; the score for each vault is the average from four judges' scores.

Suppose, over both vaults, both Smith and Jones have exactly the same marks: both get four scores of 9.65 and four scores of 9.70. A sensible scoring system would place them equal. However, suppose both of Smith's vaults are awarded two scores of 9.65, two of 9.70. Then her average score on each vault is correctly returned as 9.675, and that is also her final score.

But in her first vault, suppose Jones has three scores of 9.65 and just one of 9.70. The true average of these is 9.6625, but by scoring convention this would be rounded down to 9.662. On her second vault, the true average of her scores is 9.6875, rounded to 9.687. So when we now average 9.662 and 9.687, we get 9.6745, which is rounded to 9.674. This is below Smith. Unfair!

If, rather than averaging and rounding, the original scores were just added up, justice would be done, as both competitors would have the same total. It is indeed the rounding to three decimal places that causes the problem.

* * *

In **Chapter 5**, we mentioned a formula for the distance a shot or cannon ball would travel. Ignoring wind resistance, suppose the shot is released with speed V m/sec at an angle θ above the horizontal, from a height h metres above where it will land. Then, if the force of gravity is g m/sec/sec, the distance achieved is:

$$\frac{V^2}{2g}\sin(2\theta)(1+\sqrt{1+\frac{2gh}{V^2\sin^2(\theta)}})$$

This may look spectacular, but it is derived from standard Newtonian equations that will be familiar to any school pupil studying mechanics. The importance of the different variables can be checked by putting in some round numbers. The height of a shot putter, h, is near enough 2 m, gravity is about 10 m/sec^2 and the shot is released at about 14 m/sec at an angle of about 45 degrees. With those rounded numbers, the range is about 21.5 m. And this figure is suitably close to the world record of 23.12 m that Randy Barnes set in 1990.

What happens if these values vary? You can use the formula to assess the change in throwing distance for each 1 per cent change in the different variables.

Change in parameter	Increase in distance
1% reduction in gravity	20 cm
1% increase in height of shot putter	2 cm
1% increase in release speed	40 cm

* * *

For **Chapter 6**, we'll first remind you about logarithms to base 2. Recall that an expression like 2 x 2 x 2 x 2, i.e. 2 multiplied by itself four times, is written 2^4, and its value is 16. Turning it the other way round, the (base 2) logarithm of 16 is 4; we write $\log_2(16) = 4$. Similarly, $\log_2(32) = 5$, and since $1{,}024 = 2^{10}$, then $\log_2(1{,}024) = 10$. For a number between 16 and 32, its logarithm is between 4 and 5.

How often might we expect a run of x consecutive heads or tails somewhere in a series of tosses? Such a sequence will have a definite start point; and the chance of (at least) x consecutive heads, starting at a given point, is $(\frac{1}{2})^x$. So if there are N possible start points for such a sequence, the *average* number of times we get at least x consecutive heads will be $N \times (\frac{1}{2})^x$. For example, with 100 possible start points, we would expect around three sequences of length at least 5, because $100 \times (\frac{1}{2})^5$ is just over 3. The same argument applies to a run of at least 5 tails; on average, there will be more than six runs of 5 heads or 5 tails.

Having an average as large as 6 makes it almost certain that one or more runs of this length will occur. But the average number of times we get a run of, say, at least 10 heads, with 100 start points, will be $100 \times (\frac{1}{2})^{10}$, and this is tiny – around $\frac{1}{10}$. The same applies to tails, but even so, the average number of times we get a run of 10 or more heads or tails is only $\frac{1}{5}$. The chance of getting any such run must be quite small.

So that leads to the idea that the division point, between being *quite likely* to have a run of at least x heads or at least x tails, and this being rather *unlikely*, is when $N \times (\frac{1}{2})^x$ is a bit less than 1. And that is exactly when x is a bit more than $\log_2(N)$.

In a run of 32 tosses, there are 28 possible start points for a run of length 5. That gives an average of almost two runs (counting both heads and tails), and so some run of length 5 is quite likely. But there are only 26 start points for a run of length 7, giving an average of only 0.42 runs of this length. We are unlikely to get any such runs; the longest run is expected to be 5 or 6.

* * *

In **Chapter 7**, we suggested that, in top-class darts, the person throwing first might expect to win the leg about 65 per cent of the time. The main idea is that the best players take around 16 darts in total to reach the target but, of course, they throw groups of three darts at a time.

Discounting the possibility of finishing within an unbeatable nine darts, here are some typical chances of a good player finishing at different times:

Number of darts	10–12	13–15	16–18	19–21	22–25
Probability	10%	30%	30%	20%	10%

With those numbers, the first player will win if:

- He takes 10–12 darts – chance 10%
- He takes 13–15, and the opponent takes more than 12 – chance 30% × 90% = 27%
- He takes 16–18, opponent takes more than 15 – chance 30% × 60% = 18%
- He takes 19–21, opponent takes more than 18 – chance 20% × 30% = 6%
- He takes 22–25, opponent takes more than 21 – chance 10% × 10% = 1%

Adding these up gives a total of 62 per cent. The best players will be rather better than this, hence our estimate of 65 per cent.

* * *

One of the formulae in **Chapter 8** was to convert the chance, p, of winning a point on serve at tennis into the chance, G, of winning a service game. To win the game, you must win the last point, but the order in which the players won the previous points doesn't matter. Break down the games into the scores before that last point.

- To win in just four points, from 40–0, the chance is p^4
- From 40–15, it is $4p^3(1-p)$. $p = 4p^4(1-p)$. (Because there are four ways in which you could reach 40–15 – WWWL, WWLW, WLWW and LWWW – each with probability $p^3(1-p)$, and then multiply by p again for the final point)

- From 40–30, it is $10p^3 (1 - p)^2 . p = 10p^4 (1 - p)^2$. (Ten ways to reach 40–30)

You might also win after a deuce. Similar counting gives the result that the chance of deuce being reached is $20p^3 (1 - p)^3 = D$, say. Now let x be the chance that you win from *a score of deuce*: either you win by winning the next two points, or by sharing the next two points and winning from another deuce. This means that:

$$x = p^2 + 2p(1 - p)x$$

which leads to $x = p^2/(1 - 2p + 2p^2)$. The overall chance of winning in a game where deuce was reached is the probability of reaching deuce (D) multiplied by the probability of winning from deuce (x).

G is now found by adding up these four quantities – though it does need careful manipulation of the symbols to get:

$$G = \frac{p^4 - 16p^4(1 - p)^4}{p^4 - (1 - p)^4}$$

which is the equation stated in the chapter. (If $p = ½$, then $G = ½$, which is as it should be since each player is equally likely to win any point.)

We also noted that the order SF can never be better than FS for the two serves. The maths for this is as follows: call the chance of a fast serve being good x, and the chance of a slower serve being good y.

And let f and s be the respective chances of winning a point, assuming a fast serve is good, or a slower serve is good. We'll also make the natural assumption that a good fast serve will always be more likely to win a point than a good slower serve, i.e. that $f > s$.

If the server's strategy is FS, then he wins the point when either:

- The fast serve is not a fault, and he wins from a good fast serve, or
- The fast serve is a fault, the slower one is not, but he wins on the slower.

Together, this gives the chance of winning the point as $x.f + (1 - x)y.s$

Performing a similar breakdown for the SF strategy, the overall chance is $y.s + (1 - y).x.f$

Subtract the latter from the former: the difference is $(f - s).x.y$ But we know that f is larger than s, so this difference is always positive. Therefore the chance of winning with FS is higher than the chance of winning with SF.

* * *

We claimed in **Chapter 10** that you should expect to win about two-thirds of those soccer matches in which you scored first. The justification is as follows.

Our model, that goals come along at random, leads (by standard statistics) to the idea that the actual number of goals in a game, for a given average number, follows the so-called Poisson distribution: this leads to a formula for the chance of exactly k goals. Since we are interested only in matches with at least one goal, we modify the formula to take this into account. Let $P(k)$ be the chance of exactly k goals in such matches.

We will find the overall chance of winning if you score first, W, by breaking the matches down according to the total number of goals. In all games with just one goal, the side scoring first will obviously win. So the whole amount $P(1)$ goes towards W.

In matches with two goals, as we take the sides to be evenly matched, each side is equally likely to score the second goal. In these matches, you win half the time, you draw half the time, so the amount $P(2)/2$ goes towards W (and the same amount goes towards the overall chance of a draw).

In matches with three goals, there are four possibilities for who scores the last two (YY, YT, TY, TT, where Y means You score, and T means They score). With the first three, you win the match, with TT you lose. That means you win three-quarters of the time. Add another $3P(3)/4$ towards W.

Continue in this fashion. The final answer will depend on the average number of goals, but for the values that actually arise in soccer (i.e. somewhere between 2 and 3.5), the final answer for the chance of winning (or drawing, or losing) hardly alters, and W is close to two-thirds, as we said.

* * *

The Sonneborn-Berger method of breaking ties (in **Chapter 13**) deserves a bit more development. To illustrate, let's use the same table of results as in that chapter, but give just one point for a win. The table of results is then:

	A	B	C	D
Anna		1	1	0
Brian	0		1	1
Connie	0	0		1
Dave	1	0	0	

and the four scores are (2, 2, 1, 1), obtained by adding up the entries on a row. The S-B method then leads to (3, 2, 1, 2), and the two ties are broken – A is now ahead of B, D is ahead of C.

But there is a little snag – B and D are now equal; so to try to break this, go one step further and apply the S-B method to *these* scores. A's new 'score' is the sum of the scores for B and C, and so on. The overall answer is (3, 3, 2, 3), and it looks as though we are in trouble, because now A, B and D are all on the same score.

However, the S-B method can be repeated indefinitely until the ties are broken. The next two iterations of the process lead to (5, 5, 3, 3), then (8, 6, 3, 5) – and now every tie is broken, and we have the definite order A, B, D, C. Even better, if we continue the process, although the numbers change, *the order does not*! This leads to an unambiguous ranking that does not depend on stopping arbitrarily after one, two, three or whatever steps.

Anyone who has come across matrix multiplication will realise that we have been describing how to multiply successive powers of the original table (or matrix) by the column vector with entries (1, 1, 1, 1). Even so, it takes some university-level maths, where *eigenvalues* and *eigenvectors* are studied, to show why we will eventually get a clear order of all four players. The key name-dropping result hidden behind these computations is called the *Perron-Frobenius Theorem.*

* * *

In **Chapter 14**, we noted that even if they always beat other teams, there was a good chance that the best two teams would not both make it to the final in a knockout cup. This is because there will be N teams in the top half of the draw and N in the bottom half; one team from each half will meet in the final. If Team A is drawn in the top half, there are N–1 other places in the same half for Team B, but N places in the other half. So, with N typically 32 or 64, the chance A and B would meet before the final is 31/63 or 63/127, marginally under 50 per cent.

* * *

In **Chapter 15**, we set up a darts match, in which the winner is the person who first wins *two legs in succession.* The player throwing first wins the leg with probability p, the player throwing second wins with probability $1 - p$, call it q. The right to throw first in a leg alternates.

Listing all the sequences of wins and losses that lead to the person who throws first in the match winning according to our rule, and then adding up the chances of each possible result, eventually leads to this expression for the chance of winning if you throw first (F):

$$\frac{1 + pq^2}{(1 + p)(1 + q)}$$

The chance S that the match is won by the person who throws *second* in the first leg is found by swapping p and q in this expression. Do this, and notice that the denominators for F and S are the same, so when we subtract to find $S - F$, the numerator (all that matters) is $qp^2 - pq^2 = pq(p - q)$. Since we assume p exceeds q, S is bigger than F. Your chance of winning is higher if you allow the other person the 'privilege' of throwing first.

* * *

We now justify our claim in **Chapter 16** that, in sports where a team has N members, the number of medals should be proportional to x^N, where x is its proportion of the world's resources.

Consider the way a team is scored in a cross-country race: just add up the finishing positions of its first N runners, and the lowest score wins. Take the country's population, K, as a measure of its resources.

Take the example where $N = 4$, and the winning team has its scoring athletes in positions 5, 7, 24 and 33. There are K people who could have been 5th, then $K - 1$ who could have been 7th, $K - 2$ for 24th place and $K - 3$ for 33rd. That means there are $K(K - 1)(K - 2)(K - 3)$ different ways for that country to have had athletes in those positions. But since K, a population, is going to be pretty big, this number is well approximated by K^4.

A similar argument holds for any other value of N, and any other set of winning positions. So the resources available for a team of size N are proportional to K^N, hence its 'fair' share of team medals should follow this pattern.

* * *

Finally, let's consider the idea, in **Chapter 17**, that scoring or missing the first penalty leads to a 25 to 33 per cent difference in the chance of winning a shoot-out. The method is exactly the same as used above for computing the chance that the person who throws first should win a leg at darts.

Whether you score or miss, your opponents have five initial penalties, so we find the chances that they score exactly 0, 1, 2, . . . , 5 times. Your side will take four more penalties after the first, so the same methods give your chances of 0, 1, . . . , 4 goals among these shots. If you score from the first penalty, these are now the chances that your total score is 1, 2, . . . , 5; and if you miss the first penalty, they are the chances that your total score is 0, 1, . . . , 4.

So now, in either circumstance, the chance that you win within five penalties, or that the scores are level after five penalties, can be found. When the scores are level, each side has a 50 per cent chance of going on to win (sometime), so your overall chance of victory is known.

You have to put some numbers in and do the calculations, but the remarkable fact is that, so long as the chance of scoring from a penalty is in the range from 15 per cent to 85 per cent, the difference that scoring or missing the first shot makes is very steady, between 25 per cent and 33 per cent.

REFERENCES

Two general sources have been invaluable:

A. Chance magazine has a regular column, 'A Statistician reads the Sports Pages', written in recent years by Hal Stern and Scott Berry. These columns have been the origin of a number of the ideas discussed here.
B. The Mathematical Gazette has published, over the years, many items linking maths and sport.

We have also referred to three wide-ranging books that contain many relevant articles and explore links between maths and sport, sometimes using much deeper maths:

C. Optimal Strategies in Sports, edited by S P Ladany and R E Machol (North Holland, 1977)
D. Statistics in Sport, edited by Jay Bennett (Arnold, 1998)
E. Economics, Management and Optimization in Sports, edited by S Butenko, J Gil-Lafuente and P M Panados (Springer, 2004)

Taking Chances, John Haigh (OUP, 2nd edition, 2003) is a useful reference for general questions about probability. It also touches on some of the sports topics in this book.

For specific topics, in chapter order (some titles have been shortened):

Chapter 1

Cogwheels of the Mind. The Story of Venn Diagrams, A W F Edwards (The Johns Hopkins University Press, 2004)

Chapter 2

The Compleat Strategyst, J D Williams (Rand Corporation, 1966)

Chapter 4

'Did Lennox Lewis beat Evander Holyfield?', H K H Lee et al, *The Statistician* Vol. 51 (2002), pages 129–46

'The 2002 Winter Olympics Figure Skating', S Berry, *Chance* 15(2) (2002), pages 14–18

Chapter 5

Thank you to Weia Reinboud for permission to reproduce the graphs in this chapter.

Letter to *Nature*, A J Tatem et al (30 September 2004), page 525

www.motivate.maths.org/conferences/conf23/c23_project1.shtml for how to adjust for wind speed.

'The Shot-putter Problem', G Leversha, P Sammutt, P Woodruff, *Math Gazette*, November 1996

'The Neglected Straddle Style', M N Brearley and N J de Mestre, *Math Gazette*, July 2001

Chapter 6

'Rating Teams and Analysing Outcomes in One-day and Test Cricket', P E Allsopp and S R Clarke, *Journal of the Royal Statistical Society* 167(4), Series A (2004), pages 657–67

'Winning the Coin Toss and Home Team Advantage in One Day International Cricket Matches', B M de Silva and T B Swartz, *New Zealand Statistician* Vol. 32(2) (1977), pages 16–22

Chapter 7

'Read During Your Leisure Time', S M Berry, *Chance* 15(3) (2002), pages 48–55

'Dartboard Arrangements', G L Cohen and E Tonkes, *The Electronic Journal of Combinatorics* (2001)

Chapter 8

'Server Advantage in Tennis Matches', I M McPhee et al, *Journal of Applied Probability* Vol. 41 (2004), pages 1182–6

Chapter 9

'Using Response Surface Models for Evolutionary Estimation of Optimal Running Times', W-H Tan (in book *E* above)

'An Analysis of Decathlon Data', T F Cox and R T Dunn, *The Statistician* Vol. 51 (2002), pages 179–87

Chapter 10

'A Birth Process Model for Association Football Matches', M J Dixon and M E Robinson, *The Statistician* Vol. 47 (1998), pages 523–38

'Modelling and Forecasting Match Results', S Dobson and J Goddard (in book *E* above)

'Down to Ten', J Ridder et al, *Journal of the American Statistical Association* Vol. 89 (1994), pages 1124–7

Chapter 12

'Conversion Attempts in Rugby Football', Anthony Hughes, *Math Gazette*, December 1978

Chapter 15

'On a Theorem of G H Hardy Concerning Golf', G L Cohen, *Math Gazette*, March 2002

'Optimal Timing of Substitution Decisions', N Hirotsu and M Wright, *Journal of the Operations Research Society*, Vol. 53, pages 88–96

Chapter 16

'Inferences about Testosterone Abuse among Athletes', D A Berry and L Chastain, *Chance* 17(2) (2004), pages 5–8

Chapter 17

'On Winning the Penalty Shoot-out in Soccer', T McGarry and I Franks, *Journal of Sports Sciences*, June 2000

INDEX